Principles and Applications
of Hydrochemistry

Principles and Applications of Hydrochemistry

ERIK ERIKSSON

Department of Hydrology
Uppsala University, Sweden

LONDON   NEW YORK
CHAPMAN AND HALL

First published in 1985 by
Chapman and Hall Ltd
11 New Fetter Lane, London EC4P 4EE
Published in the USA by
Chapman and Hall
29 West 35th Street, New York NY 10001

Printed in Great Britain at the
University Press, Cambridge

ISBN 0 412 25040 3

British Library Cataloguing in Publication Data

Eriksson, Erik
  Principles and applications of hydrochemistry.
  1. Fresh water—Analysis
  I. Title
  551.48   QD142

  ISBN 0-412-25040-3

Library of Congress Cataloging in Publication Data

Eriksson, Erik, 1917–
  Principles and applications of hydrochemistry.

  Bibliography: p.
  Includes index.
  1. Water chemistry.   I. Title.
GB855.E83 1985   551.48   85-6618
ISBN 0-412-25040-3

Contents

Contents

Contents

Preface

The International Hydrological Decade (which ended in 1975) led to a revival of hydrological sciences to a degree which, seen in retrospect, is quite spectacular. This research programme had strong government support, no doubt due to an increased awareness of the role of water for prosperous development. Since water quality is an essential ingredient in almost all water use, there was also a considerable interest in hydrochemistry during the Decade. As many concepts in classical hydrology had to be revised during and after the Decade there was also a need for revising hydrochemistry to align it with modern hydrology. A considerable input of fresh knowledge was also made in the recent past by chemists, particularly geochemists, invaluable for understanding the processes of mineralization of natural waters.

With all this in mind it seems natural to try to assemble all the present knowledge of hydrochemistry into a book and integrate it with modern hydrology as far as possible, emphasizing the dynamic features of dissolved substances in natural waters. Considering the role of water in nature for transfer of substances, this integration is essential for proper understanding of processes in all related earth sciences.

The arrangement of subjects in the book is as follows. After a short introductory chapter comes a chapter on elementary chemical principles of particular use in hydrochemistry. Chapter 3, is a discussion of various hydrochemical processes following the flow path of water from the atmosphere to the soil surface, and into the domains of soils and minerals to groundwater discharge areas, where water appears on the surface in lakes and water courses. In Chapter 4, a fairly critical account of possible hydrochemical models is given and Chapter 5 is an account of environmental isotopes and their role as carriers of information on hydrological systems. The final chapter is devoted to applications of hydrochemistry and could no doubt be expanded further.

References in the text are collected and listed after each chapter. Only

Preface

references essential for the presentation are listed to avoid unnecessary burdening of the text. These lists also contain literature recommended for further study. The reader will no doubt notice that the author's name appears quite frequently in the references. This is understandable since a great deal of the ideas presented in the book appeared in one form or another during the sixties and the seventies when the author himself was involved in evaluating the role of the atmosphere in the global circulation of various substances.

Illustrations in the book are by and large original creations drawn most expertly by Mr Johan Peippo, Hydroconsult AB, Uppsala to whom the author is greatly indebted.

Erik Eriksson
Professor emeritus
Department of Hydrology
University of Uppsala

July 1984
Uppsala, Sweden

1

Introduction

Hydrochemistry is a subject which in the most general sense could cover all areas of nature which contain water and dissolved matter. However, in the present context, hydrochemistry is considered to be an integral part of hydrology and is thus somewhat restricted, excluding oceans and continental ice sheets unless, for example, their chemistry could be related to the chemistry of terrestrial waters. Perhaps it is easier to define hydrochemistry as the subject area of transformation and transportation of substances, together with the circulation of water in the continental areas of the globe, on a time scale up to a few thousand years. This description clearly excludes ocean processes and transports but not the oceans as a possible source for airborne dissolved substances.

Hydrochemistry includes parts of atmospheric chemistry since deposition of substances from the atmosphere constitutes an important source for some substances dissolved in continental waters. The subject areas of hydrochemistry and atmospheric chemistry thus partly overlap.

In nature, a considerable fraction of circulating water passes through vegetation as also do some commonly occurring chemical elements. The role of vegetation must therefore be considered in so far as it has a bearing on processes and transports of chemical substances and thus certain areas of plant physiology are relevant in this context.

The interaction between water and minerals in the ground produces additional changes in the chemical composition of the water. These interactions are caused by geochemical processes such as dissolution, hydrolysis and mineral transformations and thus, the subject areas of hydrochemistry and geochemistry also partly overlap. (Soil chemistry can be seen as a part of geochemistry.)

When groundwater emerges and accumulates on the surface of the earth in lakes and water courses, the chemical composition of the water may change again due to new equilibrium conditions of the dissolved gases.

Principles and applications of hydrochemistry

However, a frequently much more pronounced change is caused by the biological activity in the water, an activity covered by the subject area limnology, which is also of relevance to hydrochemistry.

In conclusion then, hydrochemistry, like other earth sciences, is fairly diffuse in its contours. The overlapping described can also be seen as a gradual merging of one subject area into another. There are no inherently clear boundaries anywhere. Where to draw the lines depends consequently on the context.

2

Principles of
hydrochemistry

2.1 Chemical concepts

The purpose of this section is to help in the understanding of the mechanics of atoms and molecules in the water media, basing it as far as possible on modern theories of the physics of atoms. In a sense the present picture of atomic structure is fairly classical, although the more recent application of wave mechanics to problems of interaction of atoms has been exceedingly fruitful (Pauling, 1948).

2.1.1 *Water as a chemical substance*

The water molecule is made up of two hydrogen atoms and an oxygen atom although the actual structure varies. Part of the time the hydrogen atoms form covalent bonds with the oxygen atom by sharing electrons; part of the time the oxygen atom borrows the two electrons from the hydrogen atoms, thus completing its outer shell to form the electronic structure of the inert gas neon. In this state, the oxygen atom will carry a strong negative charge matched by the positive charges at the hydrogen nuclei. This type of bond is called ionic. Further, there exist two intermediate states with one covalent and one ionic bond. There are continuous transitions between these electronic states.

The two covalent bonds in the water molecule should form an angle of 90°, but the observed angle is 105°. It is therefore thought that the partial ionic character of the bonds with both hydrogens positive, introduces a repelling force between the hydrogen atoms thereby increasing the angle by 15°. At any rate, the angle of 105° and the partly ionic character of the bonds makes the water molecule into an electric dipole usually drawn as in Fig. 2.1.

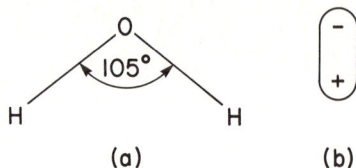

Fig. 2.1 (a) The angle between hydrogen–oxygen bonds in a water molecule; (b) the common representation of the water molecule as a dipole.

With the structure as described, the water molecule can also participate in the formation of hydrogen bonds. During the ionic state, one or both hydrogen atoms will have a positive charge. Being very small as compared with other atoms it can then attract another oxygen atom to form a relatively weak so-called hydrogen bond. Because the hydrogen ion is small it can join only two oxygen atoms at a time. In ice, these bonds aid the formation of a fairly open lattice, each oxygen atom being surrounded tetrahedrally by four oxygen atoms using hydrogen atoms as links. On melting, part of this structure collapses to give a denser packing; however, it is also thought that in the liquid state, molecular fragments of ice are continually being formed and destroyed by thermal motion. Hydrogen bonding in liquid water does, of course, affect viscosity. In liquid water the viscosity decreases about 2% per degree of increase in temperature. Hydrogen bonding in liquid water is also evident from the boiling point which is high as compared with that of methane for example, which has about the same molecular weight as water.

2.1.2 *Dissolution of chemical substances in water*

When a substance like sodium chloride is added to water it dissolves readily. At the same time the electrical conductivity of the solution increases. Pure water has a very low conductivity but when a soluble salt is added it acquires an electrical conductivity which by and large is proportional to the concentration of dissolved salt. What is happening can be described by the reaction

$$NaCl(c) \rightarrow Na^+ + Cl^-$$

where (c) stands for crystalline.

The separation of Na^+ and Cl^- can be explained as being due to the formation of clusters of water dipoles around the ions, schematically pictured in Fig. 2.2. The bond energy between the ions and the water dipoles is sufficient to keep the ions separate. They can move practically freely in the solution. There is still some attraction between the ions; the hydrated ions still cluster around each other, an important fact which forms

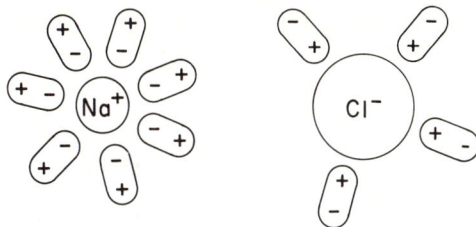

Fig. 2.2 Arrangement of dipoles of water in the vicinity of a sodium ion and a chloride ion.

the basis of the Debye–Hückel theory on the activity coefficients of ions in solutions. The density of water dipoles around an ion depends on the size of the ion and its charge. A small ion develops a stronger electric field around it than a large ion. A small ion is therefore more hydrated than a large one. Sodium is a fairly small ion, the crystal ionic radius is given as 0.95 Å while the chloride ion is fairly large, having a crystal ionic radius of 1.81 Å. Consequently, the hydration of Na^+ is much greater than that of Cl^-.

When an electric field is applied across a solution the positive ions start to move in the negative direction of the field while negative ions move in the opposite direction. The electrical 'mobility' of ions in water solutions is measured by their equivalent electrical conductance expressed in reciprocal ohms (Ω) or Siemens (S) per unit length. The equivalent conductances for Na^+ and Cl^- at 25°C are 5090 and 7550 S/m respectively. The greater hydration of Na^+ makes it slower than the Cl^- which although larger in size is less hydrated.

If another salt, for example K_2SO_4 is added to the NaCl solution it will also be dissociated into hydrated ions of K^+ and SO_4^{2-}. The solution will be a mixture of Na^+, K^+, Cl^-, and SO_4^{2-}, all of which move relatively freely in solution in their hydrated forms. There is no longer potassium sulphate in solution nor sodium chloride, just a mixture of the ions listed.

In the model for electrolytes, i.e. ions in water solution, which postulates ion associations, it is assumed that complexes like $CaCO_3$ with zero charge or $CaHCO_3^+$ exist. This approach is particularly useful to account for imperfections in the Debye–Hückel theory of strong electrolytes when applied to mixtures of dissolved salts. However, for the time being, one can regard all the common 'easily soluble salts' as completely dissociated in water solutions, the ions being hydrated to varying degrees.

When a strong acid, for example HCl, is added to water one can write the reaction

$$HCl(l) \rightarrow H^+ + Cl^-$$

5

where (l) stands for liquid. Similarly

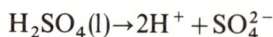

$$H_2SO_4(l) \rightarrow 2H^+ + SO_4^{2-}$$

except at very low pH values when the reaction is

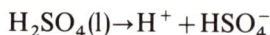

$$H_2SO_4(l) \rightarrow H^+ + HSO_4^-$$

The hydrogen ion, H^+, is a proton which associates with water molecules. Considering the previously described structure of water it is difficult to imagine it hydrated in the same sense as other cations. The equivalent electrical conductance of H^+ is very high, 35000 S m^{-1} at 25°C, thus nearly seven times greater than that of Na^+, showing its relative freedom to move.

Hydrochloric acid and sulphuric acid are said to be strong because they dissociate practically completely. There are also acids which dissociate only partly, for example, carbonic acid, H_2CO_3. It is soluble in water and dissociates in two steps. The first is

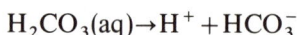

$$H_2CO_3(aq) \rightarrow H^+ + HCO_3^-$$

and the second

$$HCO_3^- \rightarrow H^+ + CO_3^{2-}$$

At the first step there are about equal concentrations of H_2CO_3 and HCO_3^- at pH=7. HCO_3^- is also seen as an acid although much weaker than H_2CO_3. In fact, H_2CO_3 includes dissolved CO_2 which is not hydrated so there are actually four species of carbon in solution.

A base is frequently defined as a substance which can consume hydrogen ions. Strong bases like sodium hydroxide and potassium hydroxide dissociate completely. As an example

$$NaOH(c) \rightarrow Na^+ + OH^-$$

A weak base is ammonia, NH_4OH which dissociates only partly into NH_4^+ and OH^-. Ammonia is, however, readily dissolved in water. Other weak bases are $Ca(OH)_2$ and $Mg(OH)_2$.

Silicic acid, H_4SiO_4, is a very weak acid. In natural waters it is found in a dissolved but undissociated state. When salts of weak acids dissolve they act as strong bases. Consider sodium carbonate: it dissolves completely in water according to

$$Na_2CO_3 \rightarrow 2Na^+ + CO_3^{2-}$$

but CO_3^{2-} reacts with water

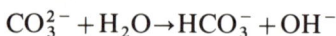

$$CO_3^{2-} + H_2O \rightarrow HCO_3^- + OH^-$$

This process is called hydrolysis. It is particularly important when silicates dissolve because silicic acid is extremely weak. For every metal ion which is released in weathering of silicates an OH^- is formed.

Summing up this section we can consider all combinations of the ions Na^+, K^+ and NH_4^+ with the ions Cl^-, SO_4^{2-}, NO_3^-, HCO_3^-, and CO_3^{2-} as easily soluble. They will dissolve under dissociation practically completely in natural waters, the exception being waters in arid environments where saturation of some or all salts may be reached. Salts of Ca^{2+} and Mg^{2+} with Cl^-, NO_3^- and SO_4^{2-} are also readily dissolved under dissociation. Under temperate conditions salts of these are not found in nature. In arid regions $CaSO_4.2H_2O$, gypsum, may form since the solubility of this salt is somewhat restricted. Combinations of Ca^{2+} and Mg^{2+} with CO_3^{2-} are not readily soluble. A possible reaction with water is

$$CaCO_3(c) + H_2O \rightarrow Ca^{2+} + HCO_3^- + OH^-$$

which increases the pH because of hydrolysis. Only small amounts can be dissolved in this way. In nature, the major dissolving agent for $CaCO_3$ and $MgCO_3$ is carbon dioxide.

2.1.3 *Ionic exchange and sorption*

Ion exchange is an important process in hydrochemistry during transient states, i.e. when changes in concentrations and ionic ratios take place. An ion exchanger is a substance which is immobile in the sense that it is not transported from a site by molecular diffusion or advection by the liquid phase. It acts as a reservoir for cations or anions. There are also ion exchangers that are amphoteric, i.e. they can store both cations and anions.

Organic matter in soils consists to a large extent of a mass of compounds which are almost impossible to identify, being residual from processes which convert organic matter in plants (and animals) to carbon dioxide and water. During this decomposition, which is largely and basically a set of oxidation processes, the organic matter acquires an acidic character to the extent that it can dissociate hydrogen ions, thus acting as an acid. The partial oxidation of hydrocarbons symbolized by (CHOH) creates –COOH groups which react like

$$-COOH \rightarrow H^+ + -COO^-$$

The acid –COOH is, however, immobile and can be pictured schematically as in Fig. 2.3. Such acids are also called acidoids. In water suspension, acidoids will dissociate part of the hydrogen as hydrogen ions. They are to some extent free to move but cannot escape because of the electrostatic attraction. They can be pictured as a diffuse swarm of hydrogen ions.

There are also inorganic crystalline substances which have similar

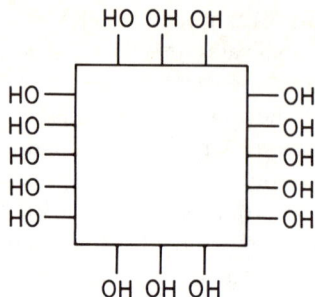

Fig. 2.3 Schematical picture of an immobile polyvalent organic acid. The OH groups can dissociate hydrogen ions.

properties. A typical one is montmorillonite, a layered mineral which because of substitution of ions in the lattice acquires negative charges which are neutralized by cations on the surface. These cations are also free to move in a diffuse swarm at the surface.

Suppose now that a sodium chloride solution is passed through the soil. The Na^+ will then replace H^+ as shown schematically in Fig. 2.4. An exchange between hydrogen ions and sodium ions has taken place. The sodium ions can also be replaced by other cations for example by Ca^{2+} if the soil is treated with a solution containing calcium ions.

An ion exchanger is a reservoir for various ions. If a solution of constant ionic composition is passed through a soil, the ion exchanger in the soil will attain a certain composition of cations. During the transient state, i.e. when the ion exchanger is adjusting itself to the composition of the solution passed through it, the percolating solution is also changing in composition. As soon as the ion exchanger attains its final composition no change in the percolating solution will be seen. Because ion exchange substances act as

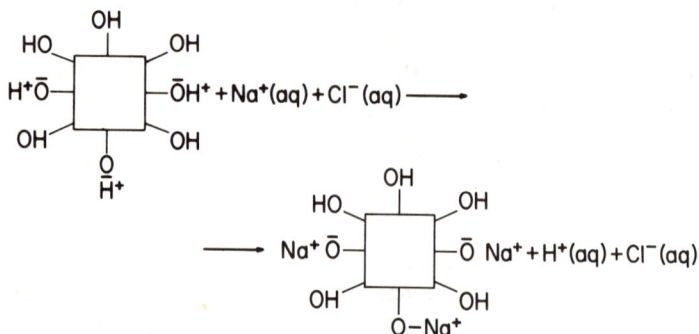

Fig. 2.4 Ion exchange, schematically pictured.

reservoirs, they will damp time fluctuations in the ionic composition of water passing through, for example, a soil. One can picture the soil–water–ion exchange complex as in Fig. 2.5.

Cation exchange materials are almost invariably present in soils. Anion exchange is rare although anion sorption is quite usual for anions of phosphoric acid, particularly $H_2PO_4^-$ and HPO_4^{2-}. Also the sulphate ion can be sorbed. The active anion sorption substances in this case are the so-called sesquioxides in soils, for example $AlO(OH)$ and $FeO(OH)$ in their microcrystalline states. Since they are almost always present, anion sorption is a common phenomena. The amounts sorbed are in general related to the concentrations of the ions in solution and are quantitatively expressed by so-called 'sorption isoterms'. Sorption creates a reservoir and this reservoir damps time fluctuations in the concentrations in solution, as in the case of ion exchange. At a steady state, sorption will have no effect on the anion composition of percolating soil water.

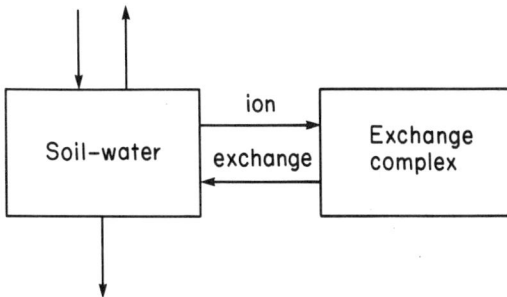

Fig. 2.5 A box model representation of ionic exchange in a soil.

Anions like Cl^- and NO_3^- are normally not affected by sorption in soils. They move along with the water and are therefore reliable tracers of water movement in soils. There are even indications that Cl^- is 'negatively' sorbed since it seems to move at a slightly higher speed than does water, as seen when using tritiated water (HTO) to trace it. The explanation for this is that the diffuse cation swarms at ion exchange sites keeps the Cl^- out of such places so that it becomes concentrated in a somewhat smaller volume than the bulk of the water.

2.1.4 *Oxidation – reduction*

Oxidation–reduction is a process whereby electrons are transferred from

Principles and applications of hydrochemistry

one substance to another. As an example, the oxidation of the ferrous iron can be written in two steps

$$Fe^{2+} \rightarrow Fe^{3+} + e^-$$

where e^- symbolizes an electron. Then

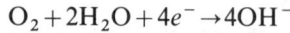

$$O_2 + 2H_2O + 4e^- \rightarrow 4OH^-$$

Ferrous iron is, hence, oxidized to ferric iron by losing an electron. Such electrons are taken up by dissolved oxygen gas which forms hydroxyl ions with water. Combining the two equations gives

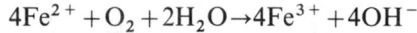

$$4Fe^{2+} + O_2 + 2H_2O \rightarrow 4Fe^{3+} + 4OH^-$$

The ferrous iron, Fe^{2+}, behaves like a bivalent ion in solution, i.e. resembles Ca^{2+} and Mg^{2+}. The ionic (crystal) radius of Fe^{3+} is less than that of Fe^{2+}. Fe^{3+} has similar properties as Al^{3+}. The reaction can only take place in strong acid solution. In natural waters the reaction would be

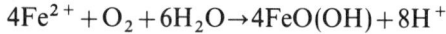

$$4Fe^{2+} + O_2 + 6H_2O \rightarrow 4FeO(OH) + 8H^+$$

for the oxidation of ferrous iron by dissolved oxygen gas. The reaction shown will almost proceed to completion provided there is some trace of oxygen gas in the water. Ferrous iron, Fe^{2+} and dissolved oxygen gas cannot exist together at equilibrium in natural waters and Fe^{3+} forms almost insoluble compounds. The small concentration found in surface water and shallow groundwater samples is mostly due to iron–humic complexes.

The reverse process, reduction, means the uptake of electrons. Hence some substances in solution must deliver these electrons. Organic substances in natural waters may be represented by CHOH. In the absence of dissolved oxygen, the reducing ability of CHOH can be illustrated by the reaction

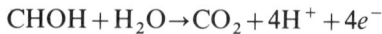

$$CHOH + H_2O \rightarrow CO_2 + 4H^+ + 4e^-$$

$$4FeO(OH) + 4e^- \rightarrow 4Fe^{2+} + 2H_2O + 6O^{2-}$$

and by adding them

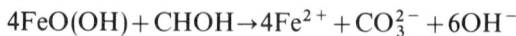

$$4FeO(OH) + CHOH \rightarrow 4Fe^{2+} + CO_3^{2-} + 6OH^-$$

The reduction process in this case produces hydroxyl ions while the oxidation produces hydrogen ions. Oxidation of ferrous iron decreases the pH while reduction of ferric iron by organic matter increases the pH.

Organic substances can also reduce nitrate and sulphate in the absence of dissolved oxygen gas. Sulphate will accept electrons, so being transformed

into hydrogen sulphide

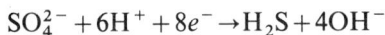

$$SO_4^{2-} + 6H^+ + 8e^- \rightarrow H_2S + 4OH^-$$

Combined with organic matter this gives

$$SO_4^{2-} + 2CHOH \rightarrow 2CO_2 + 2OH^- + H_2S$$

This seems to be a very common reaction in nature in the absence of oxygen. Nitrate, NO_3^-, can also serve as an electron acceptor under these circumstances, being converted into nitrogen gas or nitrous oxide, N_2O.

Most oxidation–reduction processes in natural waters take place with the aid of micro-organisms. These can be looked upon as catalysts although they certainly do not do this just for fun, using much of the energy released during such reactions in their metabolism.

2.1.5 *Measures and units in hydrochemistry*

The amounts of dissolved organic substances in water are normally given as concentrations, i.e. as amounts per unit volume. For natural waters the most common measure is milligrammes per litre, written $mg\,l^{-1}$. Quite frequently one finds the measure ppm, parts per million, instead of $mg\,l^{-1}$, but this should be avoided. Concentrations can also be given in moles per litre. One mole of a substance is the mass in grammes equal to the molecular weight of the substance. Water has a molecular weight of 18 so that 18 g of water constitutes one mole. Similarly, 32 g of oxygen gas constitutes one mole. For natural waters with relatively low concentrations one can use millimoles per litre. A millimole is 1/1000 of a mole often abbreviated to mmol. Other units are equivalents: one equivalent of Ca^{2+} is the atomic weight divided by two, the charge of the ion. Sodium has the atomic weight 23 so that 23 g of Na^+ is one gramequivalent. In dilute solution, like natural waters, the measure milliequivalents or even microequivalents can be used, the latter being 1/1000000 of an equivalent.

In physical chemistry $mol\,l^{-1}$ or kg^{-1} is a standard unit. Other standard units are atmospheres for gas pressures, and calories for energy. Equivalents are useful when checking the consistency of a chemical analysis of common constituents in a water sample, by comparing the sum of cations in equivalents to the sum of anions in the same measure.

Appendix A lists elements and compounds commonly occurring in natural waters, their chemical formulas, atomic weights or molecular weights and the equivalent weights of ions. When an element occurs in a chemical compound it is sometimes suitable to express the concentration of the element rather than that of the compound in which it occurs. Consider for instance nitrate, NO_3^- and ammonia, NH_3. In order to compare the concentrations with respect to nitrogen one can write $mg\,l^{-1}$ of NO_3–N and NH_3–N, nitrate nitrogen and ammonia nitrogen. As an example, 62 mg l^{-1} of NO_3 and 17 mg l^{-1} of NH_3 are 14 mg l^{-1} of NO_3–N and NH_3–N respectively. Also organically bound nitrogen can be expressed in this way by writing org–N. Similar considerations apply to phosphorous in phosphates.

Organic dissolved compounds in natural waters are as a rule impossible to identify and determine analytically by weight. In recent years quite rapid methods for determination of organic carbon in organic dissolved matter have been developed. Frequently in use is the ability of certain compounds to consume oxygen during decomposition to carbon dioxide, such as potassium permanganate, $KMnO_4$, whereby the MnO_4^- ion is converted into Mn^{2+}. One mole of $KMnO_4$ has the same oxidizing capacity as 1.25 moles of oxygen gas, O_2. In the analysis the chemical oxygen consumption (or chemical oxygen demand as it is sometimes called) is frequently given as mg l^{-1} of $KMnO_4$ (also called permanganate consumption). One millimole of $KMnO_4$ is 158 mg and has the same oxidizing capacity as $1.25 \times 32 = 40$ mg of oxygen gas. Hence, 1 mg of $KMnO_4$ corresponds to 0.25 mg of oxygen gas. It is therefore a relatively simple procedure to convert permanganate consumption into the equivalent oxygen demand.

Electrical conductivity is at present mostly expressed in milli-Siemens per metre, mS m^{-1} and is called gamma. Previously micro-Siemens per centimetre was used, called kappa. Kappa is numerically ten times larger than gamma. The electrical conductivity gives fairly accurate information on the total concentration of dissolved substances in natural waters. As a rule of the thumb the total concentration in mg l^{-1} is about 7.5 times the electrical conductivity in mS m^{-1}.

2.2 Physical chemistry applied to natural waters

In recent years a considerable number of textbooks devoted to the subject of physical chemistry have appeared and extensive treatments related to water can be found (for example, Stumm and Morgan, 1970). There is no point in duplicating this work, hence the treatment here will be restricted to those principle applications of physical chemistry which are useful for the interpretation of chemical analysis from the point of view of the origin and history of water in nature.

2.2.1 *The law of mass action*

The law of mass action is the foundation of physical chemistry. 'The rate at which a reaction takes place is proportional to the product of the concentrations of molecules taking part in the reaction.' As an example, consider the reaction

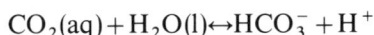

$$CO_2(aq) + H_2O(l) \leftrightarrow HCO_3^- + H^+$$

the double arrow indicating that the reaction can go both ways. The (aq) and (l) refer to aqueous and liquid states respectively. The rate at which CO_2 in aqueous solution is formed is then

$$d[CO_2]/dt = k_1[HCO_3^-][H^+]$$

where [ ] stands for concentrations. For the conversion of CO_2 into HCO_3^- and H^+ one can write

$$-d[CO_2]/dt = k_2[CO_2][H_2O]$$

At equilibrium as much CO_2 is formed as is consumed, hence

$$k_1[HCO_3^-][H^+] = k_2[CO_2][H_2O]$$

which also means that

$$[HCO_3^-][H^+]/([CO_2][H_2O]) = k_2/k_1 = K$$

We note that this embodies the definition of the law of mass action. In dilute solutions the concentration of water is practically constant and by convention it is put equal to unity. The same convention is also applied to components in the crystal (or solid) states. Then at equilibrium

$$[HCO_3^-][H^+] = K[CO_2]$$

is usually used. Now K is by no means a universal constant. It may change with temperature and it may depend on the total concentration of ions in solution. The concentration dependence can be accounted for by introducing activities instead of concentrations. The activity means active concentration and is generally expressed as a product of an activity coefficient and the concentration. Activities are usually written with parentheses instead of square brackets. Therefore the equation above should be written

$$(HCO_3^-)(H^+) = K(CO_2)$$

where K is now the thermodynamic equilibrium constant dependent only on temperature. There are still some problems since the determination of activity coefficients for the single ions is impossible in a strict sense. The way out of this dilemma is to use some approximations from the Debye–Hückel theory referred to earlier. This theory considers the interaction in a solution between positive and negative ions and its influence on the 'freedom of motion' of the single ionic species. The theory can give only rough approximations of activity coefficients. The greatest difficulty is encountered for ions in mixtures of two or more salts with ions carrying both single and double charges. For dilute solutions the activity coefficient f is expressed by

$$pf = Az^2\sqrt{I} \qquad (p \equiv -\log)$$

in its simplest form. A is a coefficient close to 0.5 which varies slowly with temperature, z is the charge of the ion concerned and I is the ionic strength of the solution as given by

$$I = 0.5\sum c_j z_j^2$$

c being the molar concentration of an ionic species with charge equal to z. More elaborate expressions are also available taking the ionic size into consideration. One which is used is

$$pf = (0.51\ z^2\sqrt{I})/(1 + 0.33\ r\sqrt{I})$$

where r is the ionic radius in solution expressed in Ångström units. The value of this parameter is given below cited from Novozamsky *et al.* (1978):

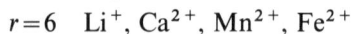

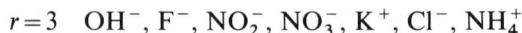

$r=9$ H^+, Al^{3+}, Fe^{3+}
$r=8$ Mg^{2+}
$r=6$ Li^+, Ca^{2+}, Mn^{2+}, Fe^{2+}
$r=5$ Sr^{2+}, Ba^{2+}, CO_3^{2-}
$r=4$ Na^+, HCO_3^-, $H_2PO_4^-$, SO_4^{2-}, HPO_4^{2-}, PO_4^{3-}
$r=3$ OH^-, F^-, NO_2^-, NO_3^-, K^+, Cl^-, NH_4^+

The data in this table may be of help in estimating activity coefficients in simpler cases. There is one peculiar inconsistency in the table. The hydrogen ion is given a large ionic radius despite the fact that the equivalent electrical conductance of the ion is extremely high compared to other ions.

Equilibrium constants derived from thermodynamic data
There is a very simple relation between the equilibrium coefficient of a reaction and the change in Gibb's free energy of the reaction ΔG_r° expressed by

$$\Delta G_r^\circ = -RT \ln K$$

where R is the gas constant and T the absolute temperature. The ΔG_r° of a reaction is obtained as the difference in standard free energy of formation from the elementary state of the products of a reaction minus the free energy of formation of the reacting components. Such data are readily found in chemical handbooks. Appendix B contains a list of ΔG_f° and ΔH_f° (enthalpies) for substances which are of interest in hydrochemistry. The use of these data can be demonstrated by a couple of examples. Consider the reaction

$$H_2O(l) \leftrightarrow H^+ + OH^-$$

From tabulated data we obtain

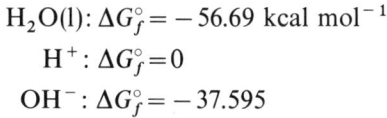

$$H_2O(l): \Delta G_f^\circ = -56.69 \text{ kcal mol}^{-1}$$
$$H^+: \Delta G_f^\circ = 0$$
$$OH^-: \Delta G_f^\circ = -37.595$$

Hence

$$\Delta G_r^\circ = -37.595 + 0 - (-56.69) = 19.095 \text{ kcal mol}^{-1}.$$

The data refer to $T = 298.15 \ K \ (25°C)$.

Then

$$RT \ln K = 1.3643 \log K = -19.095 \text{ and } \log K = -14.0.$$

Hence,

$$(H)(OH) = 1.0 \times 10^{-14}$$

As another example consider

$$CaCO_3(c) + H_2O(l) + CO_2(g) \leftrightarrow Ca^{2+} + 2HCO_3^-$$

Tabulated data give

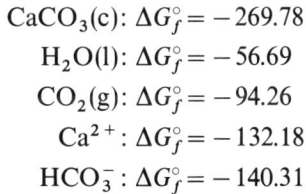

$$CaCO_3(c): \Delta G_f^\circ = -269.78$$
$$H_2O(l): \Delta G_f^\circ = -56.69$$
$$CO_2(g): \Delta G_f^\circ = -94.26$$
$$Ca^{2+}: \Delta G_f^\circ = -132.18$$
$$HCO_3^-: \Delta G_f^\circ = -140.31$$

Then

$$\Delta G_r^\circ = -132.18 - 2(140.31) - (-269.78 - 56.69 - 94.26) = 7.93 \text{ kcal mol}^{-1}$$

Hence

$$\log K = -7.93/1.3643 = -5.81 \text{ and } K = 1.55 \times 10^{-6}.$$

The equilibrium condition is therefore

$$(Ca^{2+})(HCO_3^-)^2 = PCO_2 \ 1.55 \times 10^{-6}$$

15

where PCO_2 is the partial pressure of carbon dioxide expressed in atmospheres. Since $CO_2(g)$ was used in the reaction and ΔG_f° also refers to the gaseous state carbon dioxide must be represented by its partial pressure in the equilibrium equation.

All the K values computed from standard thermodynamic data refer to a temperature of 25°C. Changes in K can, however, be worked out from the van't Hoff equation

$$\mathrm{d}(\ln K)/\mathrm{d}T = \Delta H_r^\circ/(RT)$$

Within a small temperature interval, say from 0–25°C, one can obtain a fairly good approximation of K by regarding ΔH_r° as constant. Then the equation can be integrated and yields

$$\log K_\theta = \log K_{25} - 0.733\,\Delta H_r^\circ (25 - \theta)/(273.15 + \theta)$$

As an example we consider the reaction

$$H_2O(l) \leftrightarrow H_2O(g)$$

for which $\Delta G_r^\circ = 2.055$ and $\Delta H_r^\circ = 10.522$. This gives $\log K_{25} = -1.506$ and $PH_2O = 0.03117$ atm at that temperature. Since 1 atm is equivalent to 760 mm Hg the water vapour pressure at 25°C is 23.7 mm Hg. With the value of ΔH_r°, given, one calculates that at a temperature of 10°C $\log K_{10} = -1.9148$ which gives a water vapour pressure of 9.3 mm Hg. The values given in the *Handbook of Chemistry and Physics*, edited by R. C. Weast (1975) are 23.76 and 9.25 mm Hg at 25°C and 10°C respectively.

Thermodynamic data of the kind described are extremely useful when computing equilibrium constants for reactions of immediate interest which are not readily found elsewhere in the literature.

2.2.2 The ionic product of water

Water dissociates to a small degree according to the reaction discussed earlier

$$H_2O(l) \leftrightarrow H^+ + OH^-$$

with the equilibrium condition

$$(H^+)(OH^-) = K$$

The activity of H^+ can be physically interpreted as due to jumps of protons in hydrogen bonds: A small fraction of time is spent in such jumps during which they can participate in reactions with other substances. The value of K is 1.0×10^{-14} as worked out in a previous section. In pure water dissociation of H_2O gives as much hydroxyl as hydrogen ions. The activity of hydrogen ions is therefore 1.0×10^{-7} and equal to that of hydroxyl ions. There is, obviously, a considerable degree of convention in this since it implies that the activity coefficient of the hydrogen ion is equal to the activity coefficient of the hydroxyl ion. Hydrogen ion activities are estimated by pH measurements, pH being $-\log(H^+)$. Pure water has consequently a pH of 7 at 25°C. This pH is also called the neutral point in water because the hydrogen concentration is assumed to be equal to the hydroxyl concentration at this pH. A water solution is said to be acid when pH is below 7 and alkaline when pH is above 7.

2.2.3 Gas–liquid equilibria

Oxygen and carbon dioxide are important gases in hydrochemistry. Both are present in the atmosphere, oxygen gas at a partial pressure of close to 0.2 atm and carbon dioxide at a partial pressure of about 0.0003 atm the pressures being those at sea level. The dissolution of oxygen gas in water can formally be written

$$O_2(g) \leftrightarrow O_2(aq)$$

with the equilibrium condition

$$(O_2) = KPO_2$$

The constant K depends on the temperature. Table 2.1 gives values of K at different temperatures. K decreases with increasing temperature. Considering the actual partial pressure of oxygen gas in the atmosphere

Table 2.1 The equilibrium coefficient, K, for the reaction $O_2(g) = O_2(aq)$ in the temperature range 0–20°C. (data from *Handbook of Chemistry and Physics*)

Temperature (°C)	$K \times 10^3$	Temperature (°C)	$K \times 10^3$	Temperature (°C)	$K \times 10^3$
0	2.18	9	1.74	18	1.44
1	2.12	10	1.70	19	1.41
2	2.07	11	1.66	20	1.39
3	2.02	12	1.62	21	1.36
4	1.97	13	1.59	22	1.33
5	1.92	14	1.56	23	1.31
6	1.87	15	1.52	24	1.29
7	1.83	16	1.49	25	1.25
8	1.77	17	1.47		

Table 2.2 gives equilibrium concentrations of oxygen gas in water at different temperatures, expressed in mg l^{-1}. It can be seen that temperature has a rather strong effect on the solubility of oxygen gas in water.

Table 2.2 Solubility of oxygen gas in water in equilibrium with the atmosphere at a total pressure of 1 atm and at the specified temperatures

Temperature (°C)	mg l^{-1}	Temperature (°C)	mg l^{-1}	Temperature (°C)	mg l^{-1}
0	14.55	9	11.52	18	9.45
1	14.16	10	11.25	19	9.26
2	13.77	11	10.99	20	9.08
3	13.41	12	10.74	21	8.91
4	13.06	13	10.50	22	8.73
5	12.73	14	10.28	23	8.57
6	12.40	15	10.06	24	8.41
7	12.10	16	9.84	25	8.25
8	11.80	17	9.64		

The carbon dioxide system is considerably more complicated. The reactions are as follows

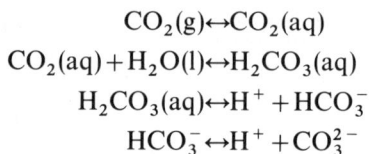

$$CO_2(g) \leftrightarrow CO_2(aq)$$
$$CO_2(aq) + H_2O(l) \leftrightarrow H_2CO_3(aq)$$
$$H_2CO_3(aq) \leftrightarrow H^+ + HCO_3^-$$
$$HCO_3^- \leftrightarrow H^+ + CO_3^{2-}$$

The hydration of carbon dioxide is not of much interest since it is very small. Therefore one usually combines the second and third equations into

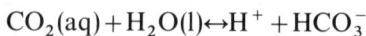

$$CO_2(aq) + H_2O(l) \leftrightarrow H^+ + HCO_3^-$$

Now the following equilibrium equations are obtained

$$(CO_2) = K_0 PCO_2$$
$$(H^+)(HCO_3^-) = K_1(CO_2)$$
$$(H^+)(CO_3^{2-}) = K_2(HCO_3^-)$$

The K values are temperature-dependent and the following relations, derived from experiments by Harned and Davis (1943) account for the temperature dependence between 0 and 50°C

$$\log K_0 = -13.417 + 2299.6/T + 0.01422T$$

$$\log K_1 = 14.8435 - 3404.71/T - 0.03279T$$
$$\log K_2 = 6.498 - 2902.39/T - 0.02379T$$

It is customary to write pK instead of $-\log K$ in analogy with pH. Table 2.3 lists some data on the pK values for selected temperatures. It is seen that pK_2 is about 10, hence $(CO_3^{2-})/(HCO_3^-) = (1.0 \times 10^{-10})/(H^+)$. In logarithmic terms this simplifies to $pCO_3 - pHCO_3 = 10 - pH$ where pCO_3 means $-\log (CO_3^{2-})$ and so on. This way of expressing equilibria will be used in the following as much as possible. In the expression above it is seen that at $pH = 10$ the carbonate and bicarbonate activities are equal. At $pH = 8$ the ratio carbonate to bicarbonate is 1 to 100. Hence, in natural waters of a pH less than 8 the carbonate can often be neglected when compared with bicarbonate.

As to bicarbonate and carbon dioxide in solution one can write

$$pHCO_3 - pCO_2 = pK_1 - pH$$

Since pK_1 is close to 6.4 the concentration of bicarbonate and carbon dioxide will be equal at about $pH = 6.4$. At $pH = 5.4$, the concentration of

Table 2.3 The negative logarithm of the equilibrium coefficients K_0, K_1, and K_2 of the carbon dioxide–water system for temperatures from 0–30°C. K_0 is the coefficent for the reaction $CO_2(g) = CO_2(aq)$, K_1 and K_2 the first and second dissociation coefficients, (data calculated from Harned & Davis, (1943))

Temperature (°C)	pK_0	pK_1	pK_2	Temperature (°C)	pK_0	pK_1	pK_2
0	1.11	6.58	10.63	16	1.35	6.41	10.42
1	1.13	6.57	10.61	17	1.37	6.40	10.41
2	1.15	6.55	10.60	18	1.38	6.40	10.40
3	1.16	6.54	10.58	19	1.39	6.39	10.39
4	1.18	6.53	10.57	20	1.40	6.38	10.38
5	1.19	6.52	10.55	21	1.42	6.38	10.37
6	1.21	6.51	10.54	22	1.43	6.37	10.36
7	1.22	6.5	10.53	23	1.44	6.36	10.35
8	1.24	6.49	10.51	24	1.45	6.36	10.34
9	1.25	6.48	10.50	25	1.46	6.35	10.33
10	1.27	6.47	10.49	26	1.48	6.35	10.32
11	1.28	6.46	10.48	27	1.49	6.34	10.31
12	1.30	6.45	10.46	28	1.50	6.34	10.30
13	1.31	6.44	10.45	29	1.51	6.33	10.30
14	1.32	6.43	10.44	30	1.52	6.33	10.29
15	1.34	6.42	10.43				

carbon dioxide is ten times higher than that of bicarbonate. In pure water there will be equal concentrations of hydrogen ions and bicarbonate ions. Since the activity coefficients of these ions in dilute solution should be the same, it follows that $pH = pHCO_3$. But $pH + pHCO_3 = pK_1 + pCO_2$ so that $pH = (pK + pCO_2)/2$. At the partial pressure of 0.0003 atm and 20°C, the $pCO_2 = 4.93$ and pK_1 can be taken to be 6.4. Therefore the pH becomes $(6.4 + 4.93)/2 = 5.66$. Increasing the partial pressure of carbon dioxide by a factor of 10 decreases the pH by 0.5 to a value of 5.16. At a partial pressure of carbon dioxide 100 times higher (not uncommon in soil air), the equilibrium pH in pure water would be 4.66.

The total amount of carbon species can be derived from the equations and becomes

$$\Sigma(CO_2) = (CO_2)(1 + K_1/(H^+) + K_1 K_2/(H^+)^2)$$

Below $pH = 8$ the last term is very small, so that in most natural waters

$$\Sigma(CO_2) \cong (CO_2)(1 + K_1/(H^+))$$

Carbonate alkalinity, usually denoted by A, is defined as 'the sum of ionic carbon species on an equivalent basis'. Hence

$$A = (CO_2)(K_1/(H^+) + 2K_1 K_2/(H^+)^2)$$

but for most purposes

$$A \cong (CO_2)K_1/(H^+) = (HCO_3^-)$$

This means that carbonate alkalinity is practically identical to bicarbonate concentration. Carbonate alkalinity is mostly expressed in meq l^{-1}. The partial pressure of carbon dioxide can be calculated from alkalinity and pH.

For other gases in the atmosphere, solubility in water follows similar rules. Expressing solubilities in the same way as for carbon dioxide the data in Table 2.4 are obtained with K in mol l^{-1} atm^{-1}. The data are worked out from the *Handbook of Chemistry and Physics*, edited by C. H. D. Hodgman *et al.*, (1956) and are presented as pK values for different temperatures. The solubilities vary considerably as seen, the high solubility of nitrous oxide, N_2O, being quite remarkable.

2.2.4 *Dissociation of weak acids*

There are a few other weak acids of interest to hydrochemists; in particular, boric acid, phosphoric acid and silicic acid. Boric acid is rather weak and dissociates according to

$$H_3BO_3(aq) \leftrightarrow H^+ + H_2BO_3^-$$

Table 2.4 Negative logarithms of equilibrium coefficients of reactions of the type $E(g) = E(aq)$ for a number of gases at selected temperatures.

Gas	Symbol	0	10	20	30
		Temperatures (°C)			
Argon	Ar	2.59	2.71	2.78	2.85
Carbon dioxide	CO_2	1.11	1.27	1.40	1.52
Helium	He	3.37	3.40	3.41	3.42
Hydrogen	H_2	3.02	3.06	3.09	3.12
Krypton	Kr	2.30	2.45	2.56	2.64
Neon	Ne	3.26	3.30	3.34	3.35
Nitrogen	N_2	2.99	3.06	3.13	3.20
Oxygen	O_2	2.66	2.77	2.84	2.92
Radon	Rn	1.64	1.83	1.97	2.10
Xenon	Xe	1.97	2.12	2.24	2.35
Nitrous oxide	N_2O	1.24	1.41	1.56	1.70

The metaform of boric acid reacts similarly

$$HBO_2(aq) \leftrightarrow H^+ + BO_2^-$$

but since BO_2^- cannot be distinguished from $H_2BO_3^-$ the first reaction is generally used. With this

$$pH + pH_2BO_3^- = pK + pH_3BO_3$$

pK being 9.22. The K value is thus small and shows that at $pH = 8.22$ only 10% of the boric acid is dissociated. In most natural waters, boron will be present as boric acid, H_3BO_3 or HBO_2. Phosphoric acid, H_3PO_4, dissociates in three steps

$$H_3PO_4(aq) \leftrightarrow H^+ + H_2PO_4^-$$
$$H_2PO_4^- \leftrightarrow H^+ + HPO_4^{2-}$$
$$HPO_4^{2-} \leftrightarrow H^+ + PO_4^{3-}$$

with the equilibrium constants (from Latimer, (1956)) $pK_1 = 2.12$, $pK_2 = 7.20$ and $pK_3 = 12$. In natural waters with a pH range of 5–8, free H_3PO_4 will be but a small fraction of total phosphorous. As to the PO_4^{3-} ion, this will not exist. The two major species will be monohydrogen phosphate, HPO_4^{2-}, and dihydrogen phosphate, $H_2PO_4^-$. At $pH = 7.2$ there will be equal concentrations of the two species.

Silica, SiO_2, forms a number of hydrates that are weak acids. The simplest is H_2SiO_3 which dissociates in two steps

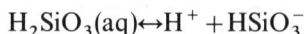

$$H_2SiO_3(aq) \leftrightarrow H^+ + HSiO_3^-$$

with $pK = 10$, and

$$HSiO_3^- \leftrightarrow H^+ + SiO_3^{2-}$$

with $pK = 12$. This demonstrates that the acid is very weak. At the pH that exists in the range of natural waters, a great number of practically undissociated polyacids of silica are formed. Sometimes they are referred to as amorphous silica with a solubility greatly exceeding that of crystalline forms, such as quartz.

2.2.5 Liquid–solid equilibria

Easily soluble salts like NaCl dissociate completely in water (at least in dilute solutions). In some instances, concentrations may become so high that NaCl forms crystals which are in equilibrium with the solution, for example in brine. There are, however, a number of minerals which are less soluble and may therefore be present at equilibrium under less extreme conditions than that of solid sodium chloride. Some of the more significant ones will be discussed below.

Gypsum

Gypsum is calcium sulphate with two water molecules attached, the formula being written $CaSO_4 \cdot 2H_2O$. The interaction with water can be written in two steps

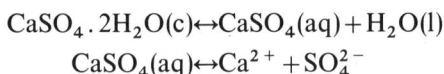

$$CaSO_4 \cdot 2H_2O(c) \leftrightarrow CaSO_4(aq) + H_2O(l)$$
$$CaSO_4(aq) \leftrightarrow Ca^{2+} + SO_4^{2-}$$

with $pK = 1.91$ for the first reaction and $pK = 2.64$ for the second reaction. The equilibrium equations are written

$$pCaSO_4 = 1.91$$
$$pCa^{2+} + pSO_4^{2-} = 2.64 + pCaSO_4$$

Calcium carbonate

Calcium carbonate equilibria are extremely important in hydrochemistry. The dissolution reaction in water can be written

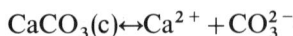

$$CaCO_3(c) \leftrightarrow Ca^{2+} + CO_3^{2-}$$

and at equilibrium when solid $CaCO_3$ is present

$$pCa + pCO_3 = pK_s$$

K_s being the solubility coefficient of calcium carbonate. When combined with the previously listed equilibria involving carbon dioxide species, the whole $Ca-CO_2$ system is set.

As to K_s this depends on the 'solubility' model adopted for $CaCO_3$. There are two ion pairs of concern which have been studied to some extent. One is $CaCO_3(aq)$, i.e. calcium carbonate molecules in solution. The other is $CaHCO_3^+$. In a study by Jacobson and Langmuir (1974) it was found that pK of $CaCO_3(aq)$ was 3.2 at 25°C and that of $CaHCO_3^+$ was 1.0 at the same temperature. They give two relations between pK_s and temperature for $CaCO_3(c)$. When ion pairs are ignored then

$$pK_s = -13.87 + 3059/T + 0.04035T$$

and when considered

$$pK_s = -13.543 + 3000/T + 0.0401T$$

where T is the absolute temperature. Since activities have to be estimated from concentrations using some form of the Debye–Hückel theory it seems safest to consider the ion pairs. One reason for introducing them is to overcome the imperfections of this theory.

There are several computer programs which have been developed for computing activities of ionic species in natural waters. All of them seem to consider specifically the ion pairs $CaSO_4(aq)$, $CaCO_3(aq)$ and $CaHCO_3^+$. A comparison between different programs published by Nordstrom *et al.* (1979) reveals some differences in the results of the ion pairs of interest. A study of their paper is recommended.

Magnesium is frequently found in calcite. A solubility product for this mixed solid (de Boer, 1977) is defined by

$$pK_x = (1-x)pCa + x \times pMg + pCO_3$$

This can be rewritten as

$$pK_x = pK_s + x \times (pMg - pCa)$$

In these expressions x is the fraction of Mg^{2+} substituted for Ca^{2+}. Thus $x = Mg/(Mg + Ca)$.

Equilibria involving two or more minerals
Mineral equilibria is a term referring to reactions by which one mineral is transformed into another. Common so-called primary minerals, i.e. feldspars, micas, hornblende to mention a few, are found in crystalline

rocks. On weathering, they are transformed into so-called clay minerals like kaolinite, montmorillonite and illite. One can also consider transformation of one clay mineral into another, for example, kaolinite into mont-morillonite or the reverse. During such transformations one may obtain highly amorphous products or microcrystalline phases which change slowly into the proper macrocrystalline phases. However, end-products of a transformation can always be predicted from thermodynamic data on the Gibb's free energies and enthalpies.

As an illustration of mineral equilibria, one can consider the transformation of soda feldspar, albite, into kaolinite in the presence of carbon dioxide. The reaction can be written

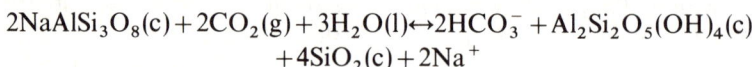

$$2NaAlSi_3O_8(c) + 2CO_2(g) + 3H_2O(l) \leftrightarrow 2HCO_3^- + Al_2Si_2O_5(OH)_4(c)$$
$$+ 4SiO_2(c) + 2Na^+$$

with the equilibrium condition

$$pNa + pHCO_3 = 0.5\ pK_c + pPCO_2$$

provided the solution is saturated with SiO_2. In the equation K_c is the equilibrium coefficient. The partial pressure of carbon dioxide comes into it as expected.

Another way to write the equation is

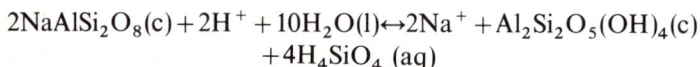

$$2NaAlSi_2O_8(c) + 2H^+ + 10H_2O(l) \leftrightarrow 2Na^+ + Al_2Si_2O_5(OH)_4(c)$$
$$+ 4H_4SiO_4\ (aq)$$

so that

$$pNa + 2pH_4SiO_4 = 0.5\ pK_c + pH$$

The choice of reaction depends on the problem at hand. In both cases thermodynamic data can be used for computing the K values.

The equations given imply equilibrium. This means that albite can be formed from kaolinite and silica provided pH or PCO_2 is in a certain range. Theoretically it is so but in practice it would not occur unless external conditions, temperature and pressure, change drastically. Not considered are all other products which can form more easily than albite at the state given.

Another reaction which may better illustrate mineral equilibria is transformation of kaolinite into gibbsite, $Al(OH)_3$.

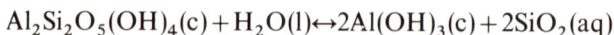

$$Al_2Si_2O_5(OH)_4(c) + H_2O(l) \leftrightarrow 2Al(OH)_3(c) + 2SiO_2(aq)$$

with the equilibrium equation

$$2pSiO_2 = pK$$

True equilibrium is established only at a specific silica concentration. When the silica concentration is higher than this value, gibbsite is transformed into kaolinite; when it is lower kaolinite is transformed into gibbsite.

Considering three minerals, say albite, kaolinite and gibbsite, a diagram of the stability of these minerals can be constructed with respect to externally controlled variables, for example, pH, pNa and $pSiO_2$. One way to do this is to use pH, pNa on one axis and $pSiO_2$ on the other. The diagrams then contain all possible equilibrium relations between any two of the minerals within permissible ranges of the variables pH, pNa and $pSiO_2$.

Kaolinite was chosen for the examples above because of its simple crystal structure. It is one of the numerous so-called layer minerals. Structurally, it consists of a layer of silica tetrahedrons joined to a sheet of aluminium octahedrons. Such Si–Al layers are then loosely stacked on each other. In montmorillonite, each aluminium octahedron sheet is joined to silica tetrahedron sheets on each side, forming a Si–Al–Si sequence of sheets. However, in the silica sheets there may be occasional substitution of Al^{3+} for Si^{4+}, resulting in excess negative charges for the crystal lattice. These charges are neutralized by cations like Na^+ or Ca^{2+} outside the lattice. Also in the aluminium octahedron sheets, there may be substitutions of Mg^{2+} for Al^{3+} also, creating charges to be neutralized by outside cations. Since the cations outside the lattice are relatively free to move bound only by electrical attraction, they will form a diffuse swarm which tends to push the Si–Al–Si layers away from each other. This is particularly evident when only the sodium ion is present. In this case the degree of dispersion of Si–Al–Si layers is very high; then they will respond to varying water vapour tensions by expanding or contracting, a phenomenon well known in soil chemistry.

When the calcium ion is the neutralizing ion, the swarm character is much suppressed and the repulsion between layers weak. Since van der Waals' forces of attraction work in the opposite direction the Si–Al–Si layers may join at edges where the electric repulsion forces are weakest. Then the layers form a kind of card house structure, very open to water and dissolved salts. Soils containing a high content of Ca montmorillonite have a favourable structure for plant available water.

2.2.6 *Oxidation–reduction equilibria*

The half reactions discussed in Section 2.1.4 can be formally treated like any other chemical reaction. Consider the reduction of ferric iron

$$Fe^{3+} + e^- \leftrightarrow Fe^{2+}$$

Treat e^-, the electron, as any other substance considered to have electron activity. As a convention, the free energy of formation of e^- is taken to be zero, consistent with the convention of setting the free energy of formation of the hydrogen ion, H^+, equal to zero. For the reaction above one obtains (see data in Appendix A)

$$\Delta G_r^\circ = -17.2 \qquad pK = -13.04$$
$$pFe^{2+} = -13.04 + pFe^{3+} + pe$$

or alternatively

$$pe = 13.04 + pFe^{2+} - pFe^{3+}$$

In this way pe is a measure of the electron activity in the solution or, one may say, the oxidation intensity. It is also related to the oxidation potential (Latimer, (1956) by the expression

$$E_h = 2.303(RT/F)pe$$

where F is 23.06 kcal and 2.303 RT at 25°C is 1.3643. Hence at 25°C

$$E_h = 0.05916 \, pe$$

When $pFe^{2+} = pFe^{3+}$ then $pe = 13.04$ and the standard potential E_h is equal to 0.771 V. Note that the half reaction is written with e^- on the left hand side in the reaction. If written in the reverse, with e^- on the right hand side then E_h becomes -0.771 V.

For water–oxygen, the half reaction

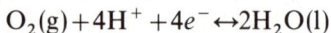

$$O_2(g) + 4H^+ + 4e^- \leftrightarrow 2H_2O(l)$$

gives $\Delta G_r^\circ = -113.38$ and $pK = -83.1$ so that

$$0 = -83.1 + pPO_2 + 4pH + 4pe$$

and

$$pe = 20.78 - pH + 0.25 \, pPO_2$$

In natural waters, the partial pressure of oxygen, PO_2, is limited to 1 atm (except in deep lakes and deep groundwater aquifers where it may theoretically exceed 1 atm). Hence a lower limit for pPO_2 is zero and

$$pe = 20.78 - pH$$

as the upper limit of pe. At this limit

$$E_h = 1.229 - 0.05916 \, pH$$

Another half reaction of interest is

$$2H^+ + 2e^- \leftrightarrow H_2(g)$$

with $\Delta G_r^\circ =$ and $pK = 0$. Hence

$$pPH_2 = 2pH + 2pe$$

or

$$pe = -pH + 0.5\, pPH_2$$

where PH_2 is the partial pressure of hydrogen. The upper limit of this should also be set to 1 atm so that

$$pe = -pH$$

and

$$E_h = -0.05916\, pH$$

For comparison, the stability of water is more directly given by the reaction

$$2H_2O(l) = 2H_2(g) + O_2(g)$$

with $\Delta G_r^\circ = 113.38$ and $pK = 83.1$. This gives

$$2pPH_2 + pPO_2 = 83.1$$

When $PO_2 = 1$ one obtains $pPH_2 = 41.55$ which means that when the partial pressure of hydrogen gas is below $10^{-41.55}$ then the water decomposes spontaneously into hydrogen gas and oxygen gas. When the partial pressure of hydrogen gas is 1 atm then PO_2 is $10^{-83.1}$ atm. For PO_2 less than $10^{-83.1}$, water will decompose spontaneously into hydrogen gas and oxygen gas.

There are now three possibilities to represent the stability region of water subjected to a total pressure of 1 atm. Water is stable within the following intervals

(a) for pPO_2 between 0 and 83.1
(b) between $pe = 20.78 - pH$ and $pe = -pH$
(c) between $E_h = 1.229 - 0.05915\, pH$ and $Eh = -0.05916\, pH$

Since pH is a variable in many oxidation reduction equilibria, it is convenient to illustrate the stability of reacting components within the limits given above. Considering pH on the horizontal axis the vertical axis can be represented by pPO_2 or pe or E_h. The stability area for water at 1 atm total pressure in these alternatives is shown in Fig. 2.6. To the author the choice of pPO_2 versus pH seems most rational but this is admittedly a matter of taste. Figure 2.7 shows the stability diagram for the system FeO–$FeO(OH)$–O_2–H_2O using the pPO_2 versus pH diagram; Fig. 2.8 shows the same using the E_h–pH diagram version.

In nature, the dominant reducing substance is organic matter. One can

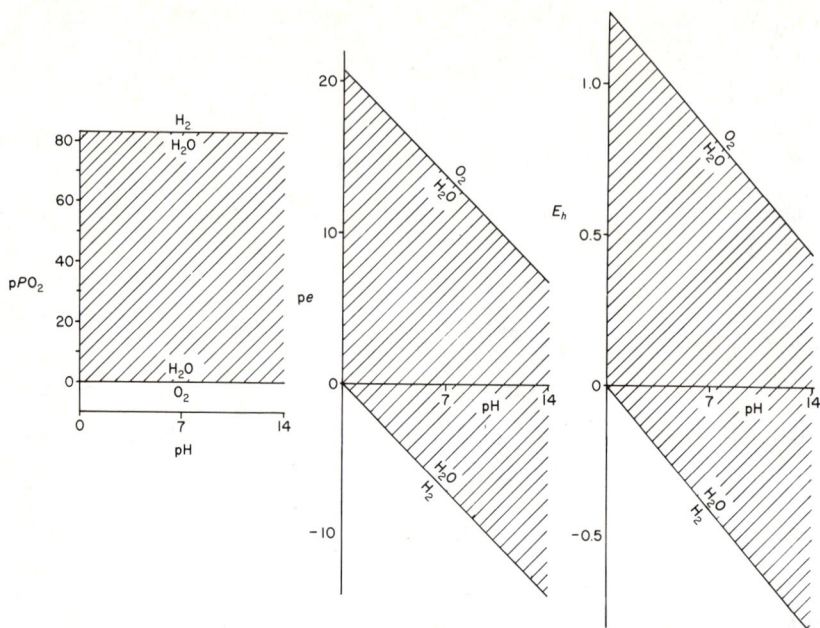

Fig. 2.6 The stability region for water in different representations of oxidation–reduction conditions. pPO_2 is the negative logarithm of the partial pressure of oxygen gas, pe the negative logarithm of the electron activity and E_h the oxidation potential.

obtain an idea of the reducing power of this substance by considering the simple reaction

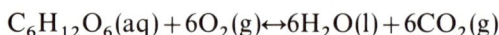

$$C_6H_{12}O_6(aq) + 6O_2(g) \leftrightarrow 6H_2O(l) + 6CO_2(g)$$

for which $\Delta G_r^{\circ} = -688.94$ and $pK = -504.98$. This gives

$$pPO_2 = 84.1 + pPCO_2 - (1/6) \times p(C_6H_{12}O_6)$$

One can conclude already from this that sugar is an extremely powerful reducing substance. If $(C_6H_{12}O_6) = 10^{-6}$ then

$$pPO_2 = 83.1 + pPCO_2$$

which means that even at this low concentration one needs a partial pressure of carbon dioxide of 1 atm to prevent decomposition of water into hydrogen gas and oxygen gas.

The example chosen is admittedly simple: in nature, breakdown of organic matter is a most complicated biochemical process. However, even

28

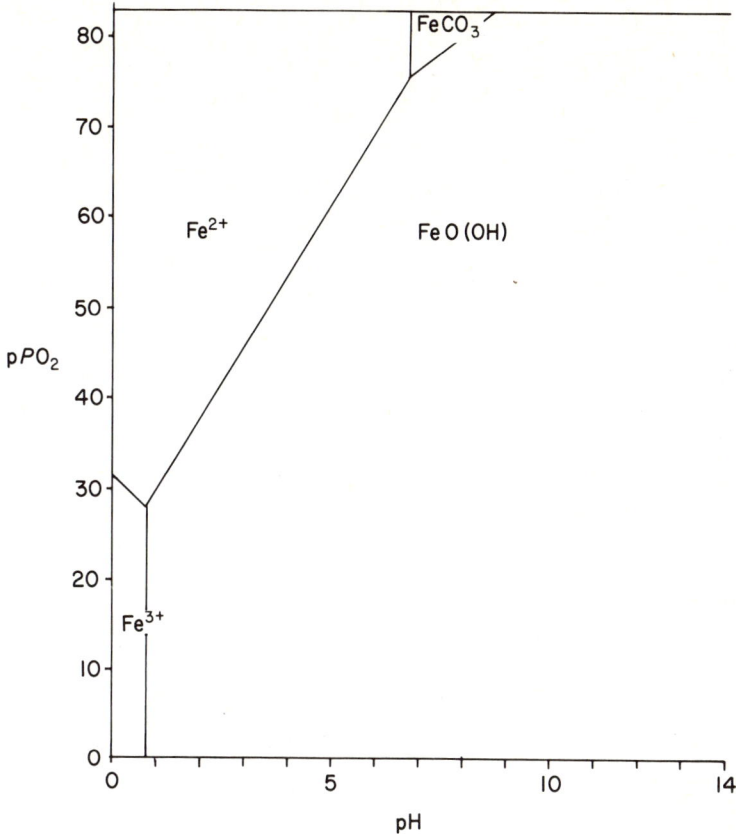

Fig. 2.7 The stability diagram for the system containing ferrous–ferric species, water, oxygen gas and carbon dioxide. Total iron 1 mmol l^{-1}, total inorganic carbon 0.1 mmol l^{-1} and total pressure 1 atm.

bacteria follow the laws of thermodynamics. It is well known that hydrogen gas is produced in bogs under conditions when water is used as a source for oxygen gas during the decomposition of organic matter.

Another reducing substance in nature is ferrous iron. Geochemically it is well known that the earth's crust is unsaturated with respect to oxygen, evident from the number of minerals containing ferrous iron, particularly in igneous rocks. Biotite is one such mineral that is relatively abundant in crystalline rocks. It decomposes by hydrolysis and by oxidation of ferrous iron into ferric iron and in this way, sets the oxidation–reduction conditions in old groundwaters which are reducing in character. This means that in these waters, all oxygen is used up and the pe is set by the Fe^{2+}–Fe^{3+} pH

29

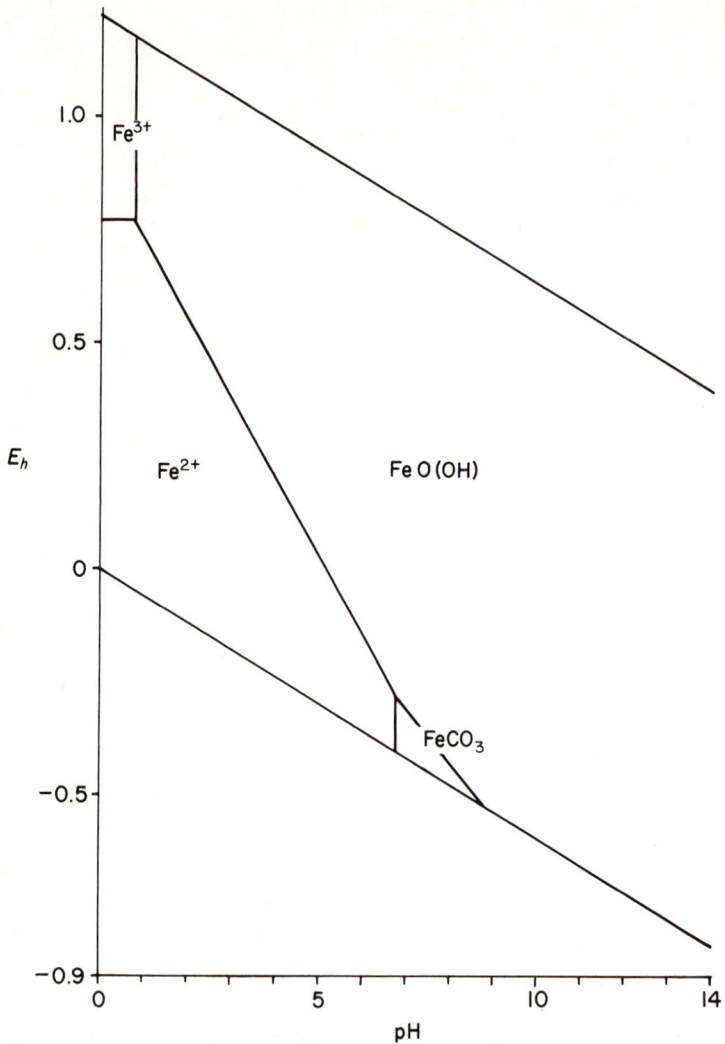

Fig. 2.8 This is the same as in Fig. 2.7 but in a E_h–pH representation.

system. In such a case, simple chemical analyses can give information about the oxidation–reduction status of the water.

References

de Boer, R. B. (1977) Stability of Mg–Ca carbonates, *Geochim. et Cosmochim. Acta,* **41**, 265–70.

Harned, H. S. and Davis, R. (1943) The ionization constants of carbonic acid in water and the solubility of carbon dioxide in water and in aqueous salt solutions from 0 to 50°C, *J. Amer. Chem. Soc.*, **65**, 2030–37.

Hodgman, C. H. D., Weast, R. C., Selby, S. M. (eds) (1956) *Handbook of Chemistry and Physics*, 38th edn, Chemical Rubber Publishing Company, Cleveland, Ohio.

Jacobson, R. L. and Langmuir, D. (1974) Dissociation constants of calcite and CaHCO from 0 to 50°C, *Geochim. et Cosmochim. Acta*, **38**, 301–18.

Latimer, W. M. (1956) *Oxidation potentials*, Prentice-Hall Inc, Englewood Cliffs N.Y.

Nordstrom, D. K., Plummer, L. N., Wigley, T. M. L., *et al.* (1979) A comparison of computerized models for equilibrium calculations in aqueous systems. *Chemical modelling in aqueous systems. Speciation, sorption, and kinetics*, (ed. E. A. Jenne), Symposium Series 93, American Chemi. Society, pp. 857–92.

Novozamsky, I., Beek, J. and Bolt, G. H. (1978) Chemical equilibria. in *Soil Chemistry A. Basic elements*, (eds Bolt and Bruggenwert), Elsevier Scientific Publishing Company, pp. 13–42.

Pauling, L. (1948) *Nature of the chemical bond*, 2nd edn, Cornell University Press, Ithaca N.Y.

Stumm, W. and Morgan, J. J. (1970) *Aquatic chemistry*, John Wiley, London.

Further reading

Garrels, R. M. and Christ, C. L. (1965) *Solutions, minerals, and equilibria*, Harper and Row Publishers, New York.

Handa, B. K. (1975) Natural waters, their geochemistry, pollution and treatment. Use of saline water, *Technical Manual No. 2*, CGWB, Ministry of Agriculture and Irrigation, New Delhi, India.

Nordstrom, D. K. and Jenne, E. A. (1977) Fluorite solubility equilibria in selected geothermal waters, *Geochim. et Cosmochim Acta*, **41**, 175–88.

Tardy, Y. and Garrels, R. M. (1974) A method of estimating the Gibbs energies of formation of layer silicates, *Geochim. et Cosmachim. Acta*, **38**, 1101–16.

Weast, R. C. (ed.) (1975) *Handbook of Chemistry and Physics*, 56th edn, Chemical Rubber Publishing Company, Cleveland, Ohio.

3

Chemical processes in
the water cycle

Features of the water cycle in nature, significant in the present context, are shown in Figs 3.1 and 3.2 in so-called box-model representations. The surface area of a basin is considered to be divided into two parts, a groundwater recharge area and a groundwater discharge area. The latter is usually found as strips along drainage channels and can vary in size depending on the basin discharge. Figure 3.1 refers to groundwater recharge areas. In both figures, the atmosphere and vegetation are represented as important parts of the water circulation. The root zone is considered as a separate entity because of its importance for water uptake by plants and for chemical and biological processes which at times can influence the chemistry of percolating water most markedly. The root zone may vary in depth from a few tenth's of a metre to several metres, depending on the plant species that grow. The intermediate zone is also called the aerated zone by hydrogeologists and the unsaturated zone by soil physicists. Below this, there is the groundwater storage zone which can extend to considerable depths; in the figures it is called the saturated zone.

The flow patterns of water are illustrated by the full drawn arrows representing precipitation, interception of water in vegetation, water uptake by roots, transpiration by vegetation and evaporation of inter-cepted water. The difference between infiltration and transpiration, flows into the intermediate zone from where it enters the groundwater storage. Yearly fluxes and storages are also given as an example, being typical for a coniferous forest site in central Sweden.

Figure 3.2 shows in a similar way the flow patterns of water in groundwater discharge areas. It is convenient to define groundwater discharge areas as 'areas where groundwater enters from below'. This means that the intermediate zone disappears. There may be an outflow of

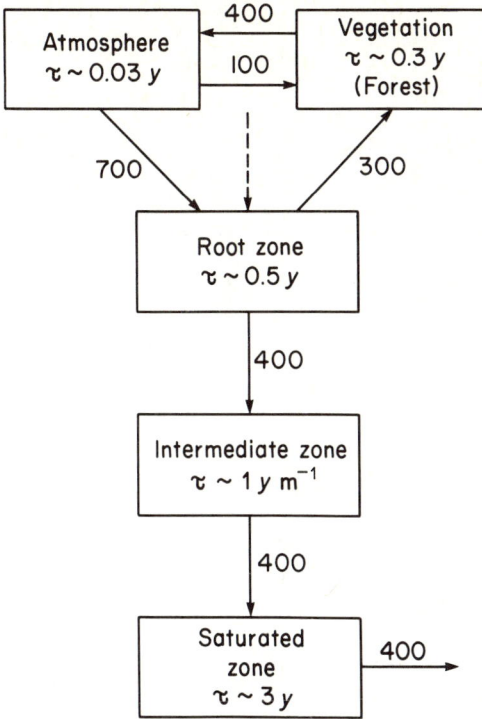

Fig. 3.1 The flow pattern of water in a groundwater recharge area in box-model representation. Turnover times, τ, typical for humid temperate climates.

water from the root zone appearing as surface run-off but this is not always the case. Evapotranspiration from the root zone will dispose of some or all of the water entering from below. Whatever is left may appear as shallow interflow, since the root zone in general has a much higher hydraulic conductivity than the intermediate zone. The flow then reaches drainage channels from where it appears as surface water, symbolized by a separate box in the figure.

The groundwater flow-rate per unit area into the root zone will vary depending on, among other things, the ratio of recharge area to discharge area. It is obvious that water availability for vegetation in groundwater discharge areas can be very much greater than in recharge areas. The ecological consequences of this are obvious.

Similar box-model representations can be made for various chemical constituents. The flow pattern may, however, differ from that of water in several respects. Such patterns and associated processes will be discussed in

33

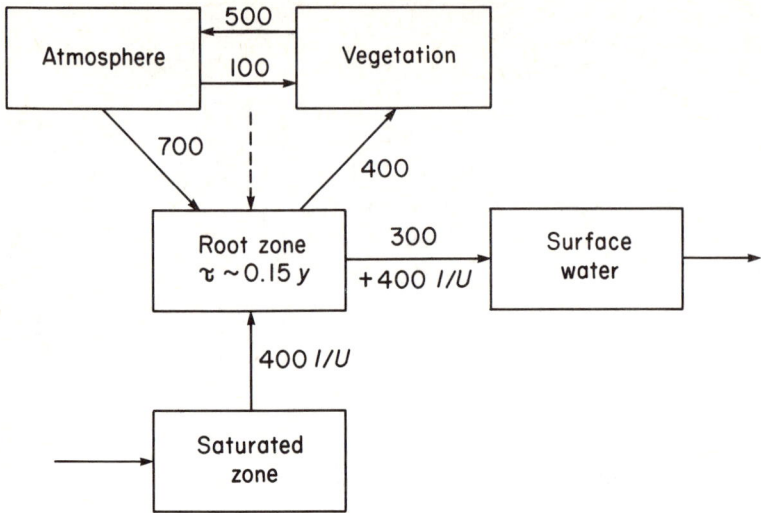

Fig. 3.2 The flow pattern of water in a groundwater discharge area in box-model representation. Turnover times, τ, typical for humid temperate climates. I/U is the ratio of recharge to discharge areas.

some detail in following sections. It seems natural to start with deposition of chemical substances from the atmosphere.

3.1 Deposition of soluble substances by rain and snow, fog and rime, and by dry deposition

The chemical composition of rain and snow was studied at a number of places from the early part of the nineteenth century, although most investigations concentrated on ammonia and nitrate, the interest being linked to the emergence of modern concepts of plant nutrition. The presence of chloride and sulphate in rain were studied in a few places also. However, more comprehensive analyses were not commenced until the early 1950s, starting in Sweden on a network basis. Gradually expansion of this network was made in Europe and other networks were established in the US and the USSR.

The early atmospheric chemistry data were reviewed by Eriksson (1952a, b). Some geochemical aspects were also discussed at length by Eriksson (1959, 1960) with the aid of more recent data. The most recent data on the chemical composition of rain and snow are being issued by the World Meteorological Organization (WMO) in collaboration with various institutions in member countries of the WMO. These data are called BAPMON data, the corresponding network being called the Global

Atmospheric Monitoring Network. Since all atmospheric stations are not necessarily background stations, there are regions with stations outside this global network. However, for the present purpose the BAPMON data are excellent so the discussion will generally be limited to these data.

3.1.1 *Origin of dissolved substances in rain and snow*

The elements that are found dissolved in rain and snow are also most common in fresh waters and are Na, K, Mg, Cl and S as sulphate. The proportions of these in rain and other fresh waters differ. There are generally speaking, two major sources of the chemical constituents in rain: one is sea salts from oceanic areas, contributing primarily Na and Cl, the other obvious source is the continental areas, particularly arid land like deserts, where Ca dominates. In addition there are so-called anthropological sources such as combustion products, accounting particularly for elements like sulphur and nitrogen in various compounds.

Rain and snow also contain particulate matter, primarily of land origin and also insoluble combustion products like soot, as well as fragments of organic matter. Some of the particulate matter may dissolve, especially in regions where output of sulphur compounds through combustion is high.

3.1.2 *Composition of precipitation*

The BAPMON data for 1978 may be considered to give a fair picture of the distribution of dissolved substances in rain and snow in the Northern Hemisphere continents, since the stations are spread fairly uniformly over North America, Europe and the temperate parts of Asia. There were 72 stations operating during 1978 of which about 60 present fairly complete analyses. A reasonably good degree of uniformity in sampling and analyses makes these data particularly valuable. Since space and time scales are interchangeable to a large extent, as far as meteorological data are concerned, the BAPMON data can be considered to give a representative picture of the deposition of chemical substances from the atmosphere.

Table 3.1 shows the statistical characteristics of the yearly means of the data presented as 10%, 50% and 90% values of the frequency distributions of the variables. A considerable part of the variability is geographic in origin, particularly that of Na and Cl with the oceans as a source. Other geographical features are displayed by the pH values which are lowest in the European region and highest on the Indian subcontinent. To a considerable degree this variable is a measure of industrial activity, i.e. of the use of fossil energy.

One special feature of the data in the table is the skewness in the

Table 3.1 Frequency distribution characteristics of precipitation chemistry from the BAPMON program for 1978. Magnitudes not exceeded in 10%, 50%, and 90% of the months

Constituent	Unit	10%	50%	90%
pH		4.2	5.1	6.5
Electrical conductivity	uS cm^{-1}	14.9	30.7	70.8
Chloride	mg l^{-1}	0.36	1.23	7.39
Ammonia nitrogen	mg l^{-1}	0.09	0.45	1.29
Nitrate nitrogen	mg l^{-1}	0.12	0.38	1.55
Sulphate sulphur	mg l^{-1}	0.37	1.21	3.29
Sodium	mg l^{-1}	0.20	0.71	7.67
Potassium	mg l^{-1}	0.10	0.28	0.93
Magnesium	mg l^{-1}	0.06	0.19	1.30
Calcium	mg l^{-1}	0.22	0.79	2.53

distribution, indicating that the distributions are near log-normal. This is supported by Fig. 3.3 which shows actual cumulative frequency distributions of the logarithms of depositions and the exceptional spread of the Cl and Na deposition rates as compared to those of K and Ca. Mg on the other hand shows a distribution fairly similar to that of Cl and Na, indicating a fair degree of marine influence.

3.1.3 *Analysis of the data*

Spatial approximately log-normal distributions of concentrations in rain and snow can be expected whenever source areas of the constituents are limited in size. For sea salt components the coastlines can be looked upon as sources, generating the salts which are then deposited at a rate proportional to the atmospheric load. This will clearly lead to exponential distributions as a function of the distance from the coastline. Similar arguments can be advanced for components like Ca, being predominantly of continental origin with source areas in arid environments. Also a compound like sulphate which is mostly of anthropological origin, is again emitted from the limited areas of intense industrial activity.

A correlation analysis of the BAPMON data on the chemistry of rain and snow can be expected to give further information about sources of the dissolved constituents. However, because of the log-normality of the distributions such an analysis has to be carried out on logarithms of the concentrations. The result of this analysis is shown by the correlation matrix in Table 3.2. Considering the number of stations used all correlation

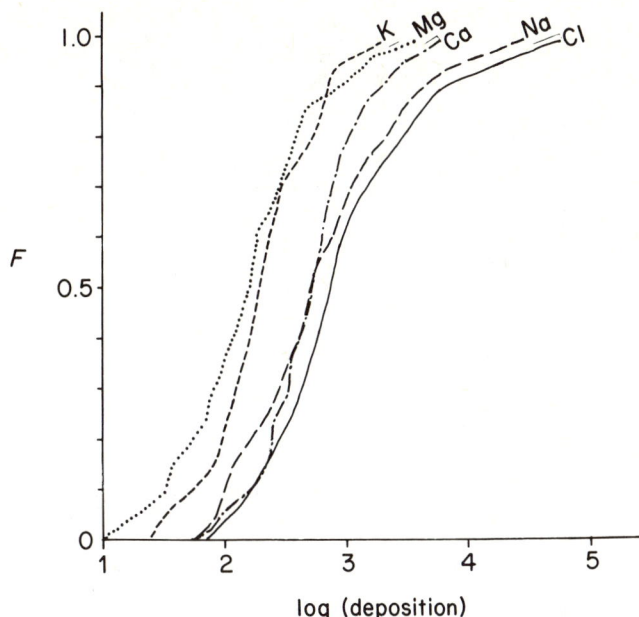

Fig. 3.3 Cumulative frequency distribution of the logarithm of wet deposition of common chemical constituents in rain and snow (BAPMON 1981).

coefficients of absolute value greater than 0.25 can be regarded as significantly differing from zero. The highest correlation 0.91, is not unexpectedly found between Cl and Na. However, also the correlation between Cl and Mg is fairly high indicating that a great deal of Mg also stems from the oceans. It is somewhat surprising to note that Cl and K are also fairly well correlated. This can be interpreted in somewhat the same way as with Mg; a fair part of K in rain and snow emanates from the oceans. About the same amounts of Ca should be of oceanic origin but the strong continental sources of Ca mask the oceanic dependence as evident from the relatively low correlation coefficient between Ca and Cl.

The elements Cl, Na, K, and Mg are thus strongly associated, forming what can be called a cluster of sea water components. Another cluster is formed by ammonia, nitrate and sulphate, as judged from the data in the table. The correlation coefficients show high affinities between ammonia and nitrate and between ammonia and sulphate while the association between nitrate and sulphate is somewhat weaker. No doubt the reason for this is the release of nitrogen and sulphur compounds during combustion. Ammonia and sulphur are emitted during combustion, the latter appearing as sulphur dioxide, SO_2, which is oxidized to sulphate in the atmosphere or

Table 3.2 Correlation matrix from the 1978 BAPMON data*

	1	2	3	4	5	6	7	8	9	10	11
1	1.00	-0.11	0.02	-0.33	-0.11	-0.39	0.02	0.25	0.26	0.30	0.01
2	-0.11	1.00	0.71	0.19	0.27	0.42	0.64	0.42	0.59	0.43	0.02
3	0.02	0.71	1.00	-0.18	-0.14	0.17	0.91	0.63	0.76	0.38	0.19
4	-0.33	0.19	-0.18	1.00	0.66	0.62	-0.15	0.18	-0.09	0.34	-0.17
5	-0.11	0.27	-0.14	0.66	1.00	0.43	-0.06	-0.08	-0.15	0.13	-0.09
6	-0.39	0.42	0.17	0.62	0.43	1.00	0.23	0.25	0.21	0.22	-0.19
7	0.02	0.64	0.91	-0.15	-0.06	0.23	1.00	0.66	0.76	0.42	0.10
8	0.25	0.42	0.63	0.18	-0.08	0.25	0.66	1.00	0.63	0.68	-0.08
9	0.26	0.59	0.76	-0.09	-0.15	0.21	0.76	0.63	1.00	0.61	-0.02
10	0.30	0.43	0.38	0.34	0.13	0.22	0.42	0.68	0.61	1.00	-0.19
11	0.01	0.02	0.19	-0.17	-0.09	-0.19	0.10	-0.08	-0.02	-0.19	1.00

*The numbers refer to the following constituents:

1 pH	4 NH_3–N	7 Na	10 Ca
2 Electrical conductivity	5 NO_3–N	8 K	11 Precipitation
3 Cl	6 SO_4–S	9 Mg	

when absorbed by soil and vegetation. The origin of nitrate is somewhat obscure. It may be formed by oxidation of ammonia in the atmosphere or during combustion but it can also be 'fixed' from atmospheric nitrogen gas due to strong heating and rapid cooling of air in stack gases. This is known to produce nitric oxide, NO, which is gradually converted into nitric acid by oxidation. This possibility is strongly supported by the fact that the nitrate fraction of inorganic nitrogen compounds found in rain and snow has increased considerably after the last world war. In the data cited by Eriksson (1952a, b) the ratio of ammonia nitrogen to nitrate nitrogen was about 2 up to 1940, while the present ratio is close to 1. More efficient combustion of fossil carbon would mean higher temperatures and better conditions for the formation of nitric oxide.

It can also be seen from the table that there is a relatively strong negative correlation between the pH and the logarithm of ammonia and sulphate concentrations. The pH gives a fair measure of the acidity in rain and snow attributable by and large to sulphuric acid from coal and oil, the ammonia being largely of the same origin. Although ammonia will neutralize some of the sulphuric acid formed there will always be an excess of the acid.

The results of the correlation analysis can be schematically depicted as in Fig. 3.4, using a cluster description. From this, the strong association

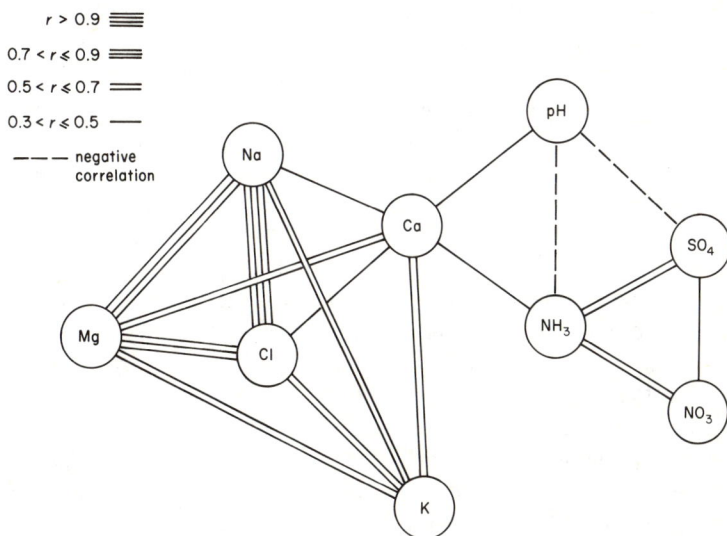

Fig. 3.4 Cluster description of the BAPMON rainfall chemistry data (BAPMON, 1981). The strength of correlation is given by the number of lines joining each pair. For $r > 0.9$ there are four lines, for $0.7 < r < 0.9$, three lines, for $0.5 < r < 0.7$, two lines and for $0.3 < r < 0.5$, one line. Dashed lines represent negative values of r.

between oceanic elements is clearly seen as well as the somewhat less pronounced cluster of ammonia, nitrate and sulphate. A weak link between the two clusters is formed through calcium, the origin of which is not only oceans and deserts but presumably also agricultural areas where a high lime status of the soils is maintained.

3.1.4 Wet deposition

The deposition of soluble substances in rain and snow is obtained by multiplying the concentrations with the amounts of rainfall/snowfall. A suitable deposition unit is mg m^{-2} year^{-1} which is numerically the same as kg km^{-2} year^{-1}. Note that 1 mm of rain is 1 l m^{-2}.

Deposition rates were worked out from the arithmetic means of concentrations and the listed rainfall figures. The latter are, in a number of cases, somewhat less than rain and snow values measured by standard rain gauges, being calculated from the volumes of collected samples. The deposition data are therefore minimum figures. Using yearly mean concentrations and yearly precipitation figures may introduce another error when concentrations and monthly precipitation amounts are correlated. For a negative correlation such calculated depositions will be overestimated; when correlations are positive they are underestimated. However, the error is not particularly large and is unimportant in the present context.

The deposition data are presented in the same way as the concentration data, i.e. as the 10%, 50% and 90% points on the frequency distributions presented in Table 3.3. The median, i.e. the 50% points can be taken as representative deposition rates in areas of medium population density at some distance from the coastlines and under temperate climatic conditions. The 10% points represent more extreme conditions such as high altitudes, sparse population, large distance from coastlines. The 90% points represent coastal regions densely populated and industrialized. The 90% and 10% data cannot be taken as typical for relations between the deposition rates of the various components since they represent areas which are extreme in one way or another.

Practically all the BAPMON data are from the Northern hemisphere. Conditions in the Southern hemisphere are less well known. In Australia rainfall was collected at 29 stations for chemical analysis in the years between 1950 and 1956 (Hutton and Leslie, 1958; Hingston, 1958). Their data on sea salt components are of particular interest and are shown in Table 3.4 as 10%, 50% and 90% values on the frequency distributions of yearly depositions. Comparing these data with those in Table 3.3 it is seen that the 10% data for Australia are somewhat higher than those from the

Table 3.3 Frequency distribution characteristics of wet deposition obtained from the BAPMON program during 1978. Depositions not exceeded at 10%, 50% and 90% of the stations

Constituent	Depositions ($mg\ m^{-2}\ year^{-1}$)		
	10%	*50%*	*90%*
Chloride	165	750	6000
Ammonia nitrogen	65	300	1040
Nitrate nitrogen	80	310	1300
Sulphate sulphur	240	970	2800
Sodium	100	510	5000
Potassium	50	200	750
Magnesium	35	160	1000
Calcium	180	510	2200

Table 3.4 Frequency distribution characteristics of wet deposition at 29 Australian stations in the years 1950–56. Depositions not exceeded at 10%, 50% and 90% of the stations (Data from Eriksson (1959))

Constituent	Depositions ($mg\ m^{-2}\ year^{-1}$)		
	10%	*50%*	*90%*
Chloride	400	1100	9400
Sodium	400	1000	5000
Potassium	50	100	300
Magnesium	100	200	800
Calcium	200	300	1000

Northern hemisphere. This is understandable since the Australian stations operated only in Victoria and Western Australia, the arid regions not being represented. The median deposition are for the same reasons higher than those in the northern hemisphere. The 90% data in Table 3.4 represent stations close to the coast.

In Africa there are data from at least four stations in 1958 (Eriksson

Principles and applications of hydrochemistry

(1966)). The wet depositions are listed in Table 3.5. Three of the stations are situated in the former Belgian Congo and are representative of the humid tropical climate of the equatorial Africa. The fourth station, Pretoria in South Africa, may be taken to represent the semi-arid parts of southern Africa, perhaps also the East African highlands. The data are not much different from the median of the Northern hemisphere data.

The information on the chemistry of rain and snow from various parts of the globe seems to make it possible to estimate wet deposition rates in various parts of the world with a fair degree of reliability.

Table 3.5 Wet deposition of different constituents at four places in inland Africa (data from Eriksson (1966))

Station	Cl	NH_3–N	NO_3–N	SO_4–S	Na	K	Mg	Ca
				Depositions $(mg\ m^{-2}\ year^{-1})$				
Yangambi*	400	150	460	700	210	210	120	370
Mulunga*	400	530	420	450	160	260	110	340
Binza*	220	70	290	280	130	140	90	460
Pretoria	390	80	160	1000	270	180	390	1440

*The first three stations are in former Belgian Congo.

3.1.5 Deposition by fog and rime

Wet deposition can also take place through fog and rime. The concentrations reported for fog water are in general very high as compared to those in rain water (Eriksson, 1959, 1960). However, fogs are rather localized and can hardly contribute to the total deposition except where rainfall is low and the frequency of fog high. Fogs are frequent on coasts where up-welling of cold waters take place and also at high altitudes (being actually clouds). A great deal of water is swept out of such fogs by trees. It consequently adds to the deposition.

3.1.6 Dry deposition

Dry deposition is defined as 'the flux of chemical compounds in the absence

42

of rain or snow or fog to the ground'. Dry deposition of sea salt particles and other easily soluble particles in air (condensation nuclei) is facilitated by vegetation which literally sweeps the particles from moving air, the particles impingeing on leaves, twigs and branches. The fall velocity of condensation nuclei in air is mostly far too low to be of any importance for dry fall-out. Hence, impingement seems to be the only possible process.

Dry deposition of gases like ammonia, nitric oxides and sulphur dioxide is facilitated when the ground acts as a sink for these gases. Ammonia may be absorbed by the soils through the acidic properties of decaying organic matter. Nitric oxides (often written NO) are highly reactive and sulphur dioxide can certainly be oxidized to sulphuric acid on contact with soil and vegetation cover. In such cases the factor that determines the deposition rate, is the concentration in air and the rate of exchange of air between 'free air' and air in close 'contact' with the soil and vegetation. This exchange can be quantitatively described as a deposition velocity. Considering 'free' air at 2 m above the soil or vegetation surface it has been found experimentally that the deposition velocity is of the order $cm\ s^{-1}$ and an often cited value is $2\ cm\ s^{-1}$. However, considering the variability of properties of particles and gases and of soils and vegetation deposition velocities ranging from $0–2\ cm\ s^{-1}$ seem reasonable. Eriksson (1959, 1960) comparing wet deposition of chloride with river run-off of chloride in Sweden and in Finland, accounted for the difference as being due to dry deposition of sea salt particles. Since chloride in soils and rocks in this area can be entirely neglected, the dry deposition rate of chloride could be worked out from the balance, and the concentrations of chloride in air samples taken at the same time as the precipitation samples. The calculated dry deposition rates at 28 stations varied from $0.2–3.5\ cm\ s^{-1}$.

Dry deposition can thus be estimated from measurements on the concentrations in air samples. However, dry deposition of chloride may be obtained from chloride balance studies which is feasible in most areas. Quantitatively, the dry deposition may account for as much as the wet deposition but may in some cases be insignificant.

3.1.7 *Exudation by vegetation*

Precipitation collected under trees, for example spruce, may contain comparatively high potassium concentrations indicating exudation of potassium. Also other ions may show similar features, at least cations. Whatever takes place, exudation of ions can be considered to be a part of the cycling between plant and soil. It will not necessarily add anything new to the larger cycle and to the budget.

3.1.8 *Deposition through leaching and decay of litter*

Litter in the present context consists of organic residues with well-recognizable structures. In undisturbed environments the chemical elements and compounds transported from the soil into the plant during its growth are mostly returned to the soil surface by the litter. There, it is leached and decomposed releasing all its mineral constituents which are thus 'deposited' onto the soil surface. Quantitatively this deposition is appreciable. Table 3.6 gives some data on such deposition rates in a few localities with different forest vegetation. The figures are in general much higher than those of wet deposition except maybe for sulphur which is comparable in size. Calcium, potassium and nitrogen are particularly high. Hence, the local cycle between soil and plants (at least for those elements which are more or less vital to the plants) is almost one order of magnitude larger than the deposition of airborne constituents.

Table 3.6 Deposition of mineral substances in litter fall at five places in Scandinavia (data cited by Troedsson and Nykvist (1973))

Vegetation	Site	Depositions $(mg\ m^{-2}\ year^{-1})$				
		Ca	Mg	K	S	N
Spruce	Ås, Norway	3600	–	900	–	3900
Spruce	Velda, Norway	3200	–	400	–	2200
Birch	Velda, Norway	2900	–	500	–	2000
Spruce	Kongalund, Sweden	2000	300	1100	500	5800
Beech	Kongalund, Sweden	3200	400	1400	600	6900

3.1.9 *Effects of human activity on deposition of chemical constituents*

Human activity can have two effects. One is the import of minerals into the basin, distributing this over the area onto the ground or surface waters. An example of this is the import of sodium chloride to a densely populated rural area. One such case is well described by Jacks and Sharma (1982) in a study of the salt balance of the Noyil River Basin in South India. Also the import of fertilizers and their use in agriculture and forestry increases deposition rates of some substances. Liming of agricultural land is a common practice in humid regions and increases deposition of calcium.

The other effect of human activity causes decrease in deposition because of removal of harvest products in agriculture and forestry. This is, geochemically speaking, an interesting process which started when man gave up hunting and nomadic life and diversified in occupation. The growth of settlement centres, villages and towns, caused a flow of products from cultivated and wild lands into the centres, depriving the producing areas of chemical constituents which were released on a concentrated scale in the communities increasing the load of chemicals in local groundwaters and surface streams. This process is, of course, still going on although it is balanced by fertilizer use at least in agriculture, the fertilizers being taken from geological storage in rocks or from the atmospheric nitrogen.

There is thus a considerable impact of human activity in densely populated areas. In Jacks' (1963) study from South India of the sodium chloride balance, the human consumption of this salt being imported, was three times the estimated total deposition of sodium chloride from the atmosphere. The population density in this case is rather high, 310 km^{-2} and a large part of the population consists of farmers.

In sparsely populated areas the effect of human activity on the deposition of chemical constituents will be small and can be neglected in balance studies. There is of course a human activity part in wet and dry deposition but this is measurable and already taken into account. This human contribution is certainly significant in other respects.

3.2 Processes in the root zone in groundwater recharge areas

3.2.1 *The root zone*

From a hydrological point of view the root zone is the part of the ground from which plant roots are able to extract water during transpiration. From a pedological point of view, the root zone consists of the so-called A and B horizons. The A horizon is the upper part, including litter and humus, having a greyish appearance while the B horizon frequently has a yellowish to red colour. Both horizons show fragments of plant residues, especially of roots, as well as the dark-coloured organic matter usually termed humus.

There is a great variety of soil types in the pedological sense. One very widespread type in humid temperate climates is the podsol, with a sharp delineation between the A and B horizons. The A horizon is mostly made up of two distinct layers, one consisting of organic residues, raw humus or moor, below which a layer of ash grey mineral soil is found. The B horizon is yellow coloured, sometimes darkened by organic matter. The yellow colour gradually thins out and disappears with increasing depth at what is supposed to be the end of the B horizon and the beginning of the C horizon, generally regarded as the parent material of the other two. Hydrologically, the C horizon should be the beginning of the intermediate zone.

In tropical humid climates, the A horizon is hardly noticeable except from the litter which may be accumulated. The reason for this is that the decomposition of organic matter is very rapid and the soil fauna mix the upper part very efficiently. These soils are mostly red in colour.

The soil types just described are typical for groundwater recharge areas. Thus, weathering products move downwards with percolating water, some out of the root zone, others (as in the case of podsols) only a small distance before being precipitated. This transport process, also called leaching, is an essential and characteristic process in groundwater recharge areas. The actual weathering processes particularly on primary minerals like feldspars, release sodium, potassium, magnesium and calcium together with amorphous silica. These constituents are then leached, i.e. transported with the moving water. The residue of weathering will be clay minerals primarily ferric and aluminium oxides. The latter may combine with silica to form kaolinite if the leaching rate is not too high. Potassium feldspars may weather into illites whereby part of the potassium will be stored but again this depends on the availability of silica.

Present weathering, i.e. mineral transformation rates in root zones, are rather low and do not contribute much to the amounts of dissolved salts. However, the root zone is an important place for the turnover of organic matter which is perhaps the most important part of the ion exchange reservoir of root zones. The exchangeable ions in the root zone can be looked upon as a storage. During a steady state flow of ions through the root zone the ion exchange material has no function since it does not influence the chemical composition of the water. If there are fluctuations in the ratios of ions being deposited onto the ground then the ion exchanger acts as a filter by damping the fluctuations during the passage through the root zone. However, if there is a sudden change in exchange capacity due to changes in organic matter content this will have a great effect on the chemical composition of the percolating water.

Clay minerals in the root zone can also act as ion exchangers. Ferric and aluminium hydroxides have, however, very low capacity for ion exchange and the same is true for kaolinite. Illites are somewhat better in this respect. Ferric and aluminium hydroxides (of the $FeO(OH)$ type) may attract anions, particularly sulphate in the prevailing pH range. At high pH values, however, they change into cation exchangers. Their importance for storing sulphate in soils is not well known. With aluminium oxides sulphate may form minerals. At low pH values one such mineral is $AlOHSO_4$. Also phosphate reacts with the ferric and aluminium hydroxides and seems to form rather well-defined compounds.

The ferric and aluminium hydroxides associate readily with organic matter in the soil and are important in so far as they prevent the transport of

humic substances with percolating water. These adsorbed humic substances may still behave as acids and ion exchangers.

The organic matter production by decaying roots is a major process
matched by the rate of decomposition of organic matter which is carried out
by a fairly well diversified microflora and microfauna. Nitrogen and
sulphur are important ingredients in this process. The major storage area of
these elements is in the soil organic matter.

3.2.2 *Effects of evapotranspiration on dissolved salts*

Deposited dissolved salts which are transported by infiltrating water into
the root zone will be concentrated by evapotranspiration. In vegetation
covered areas, the abstraction of water from the soil is almost entirely by
root uptake and transpiration of plants. During this process also some
cations and anions may be taken up by plant roots and transferred into the
green parts of the plant. The chloride ion does not seem to be much involved
in root uptake except by certain specialized plants in arid environments.
Hence, chloride concentrations will vary roughly inversely with the water
content of the soil. Also the sulphate ion can be expected to behave
somewhat similarly to chloride although plants will take up some. Nitrate is
prone to root uptake and its concentration will therefore be strongly
affected by this. The common cations Na, K, Mg and Ca in soil solution are
in most cases in exchange equilibrium with fairly large storages of
exchangeable ions. When the water content of the soil decreases the total
concentration of these cations also increases proportionately since they
must match the anions present. However, the ratios between cation species
can possibly be affected by increasing concentrations. But since the storage
of exchangeable ions is large – often more than 100 times the storage in
solution – the ratios of concentration of cation species in the soil solution is
at any moment determined by the storage of exchangeable ions. This
storage, however, reflects the long term average of deposited cations and
cations released by weathering as discussed earlier.

In a system where no abstraction of plant material is made, the ions taken
up by roots will be returned to the soil. There will be no loss except perhaps
for nitrogen and sulphur which can escape (or be added) in the gaseous
state. On a long term basis one can relate C, the concentration of common
ions in the soil solution to the deposition rates D, precipitation intensities
P, evapotranspiration rates E, release by weathering W, and escape of
volatile compounds V by the general expression

$$C(P - E) = D + W - V$$

47

For chloride we would have

$$(Cl) = D/(P - E)$$

since release by weathering is practically nil.
For cations

$$(M) = (D + W)/(P - E)$$

where (M) stands for any cation under consideration except ammonia. For sulphur and bound nitrogen, symbolized by A

$$(A) = (D - V)/(P - E)$$

since weathering hardly contributes to sulphur, at least in the root zone.

All the expressions refer to stationary states in the statistical sense, i.e. fluctuations are permitted but no trends. This is probably well-fulfilled for chloride and to a reasonable degree for cations although changes in organic matter content due to changing land-use practices can have strong temporary effects also on sulphur and nitrogen which are stored in organic matter in considerable quantities.

It can be seen from the equations that concentrations become very high when evapotranspiration almost equals precipitation. This, of course, happens in arid areas and can lead to crystallization of various compounds. Considering the solubility of compounds in which common cations can occur, calcium carbonate seems to be the first one to precipitate as calcite (or aragonite which is often a precursor to calcite). Also magnesium carbonate, brucite, and the mixture of calcite and brucite called dolomite may form. The precipitation of calcite leads to cementation of soil particles and possibly to the formation of hardpans. Considering deposition of atmospheric constituents, an arid climate would trap the calcium as carbonate, the rest of the airborne products being pushed as a strongly concentrated salt solution down into the intermediate zone. This situation can be illustrated with data from an investigation by Foster *et al.* (1982) carried out in the Botswana part of the Kalahari in an area where the rocks are covered by the Kalahari beds, sandy, wind-transported deposits. They sampled the unsaturated zone down to a depth of 20–30 m. They washed out the soluble salts, discovering, among other things, the presence of chloride. From the volumetric water content they were able to work out the quantity of chloride at varying depths. Figure 3.5 shows profiles of chloride, its concentration and accumulated amounts. The pattern can be interpreted in terms of deposition rates of chloride and recharge rates of groundwater, the latter being very low. At a few metres depth, the sand became partially cemented indicating formation of calcium carbonate. The

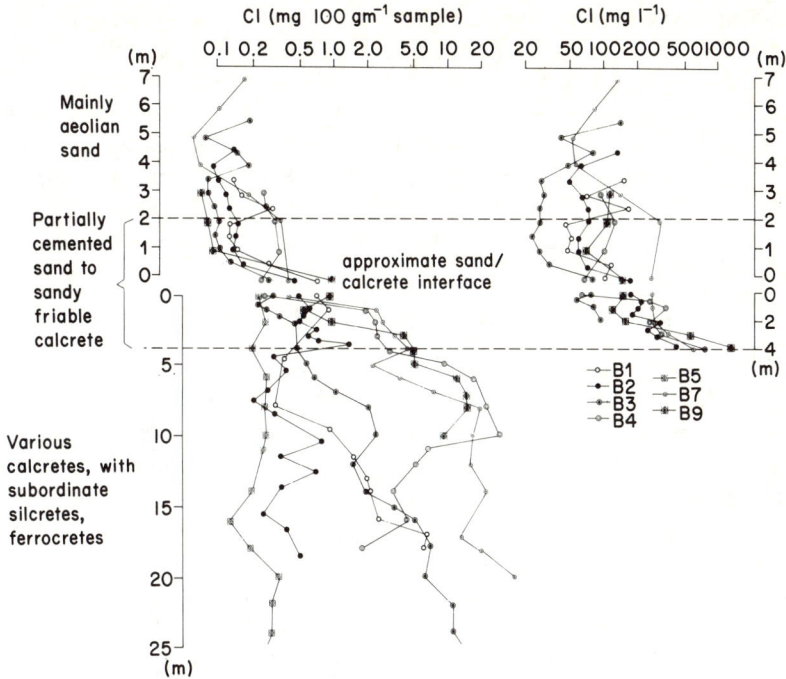

Fig. 3.5 Concentrations in water and amounts of chloride in profiles through the Kalahari beds in Botswana (data from Foster *et al.* (1982), reproduced by permission from Elsevier Scientific Publishing Company).

area where the investigation was carried out can be termed arid in the proper sense of the word.

3.2.3 *The oxygen and carbon dioxide exchange, atmosphere–soil*

Production of organic matter by plants seems to be intimately tied to the transpiration of plants. A simple rule says that in order to synthesize 1 kg of organic matter through assimilation of carbon dioxide, the plant has to transpire 400–500 kg of water. Since evapotranspiration (except in desert areas) varies from 200–1000 mm yr^{-1} (i.e. 200–1000 kg m^{-2} yr^{-1}) the production of organic matter per year should vary between 0.5 and 2.5 kg m^{-2}. About half of this production will go to the root system. To balance, the same amount must be converted to carbon dioxide and water in the root zone with the aid of oxygen.

Hence, according to the reaction

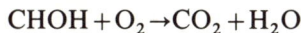

$$CHOH + O_2 \rightarrow CO_2 + H_2O$$

49

Principles and applications of hydrochemistry

1000 g of organic substance, CHOH, reacts with 1067 g of oxygen to form 1467 g of carbon dioxide and 600 g of water per m² of soil per year, choosing a production figure of 2 kg m^{-2} yr^{-1}. This means that about 750 litres of pure oxygen gas must enter the soil and 750 litres of carbon dioxide must leave the soil by molecular diffusion, all expressed per m² per year. The effect of this on the partial pressures of carbon dioxide and oxygen gas will now be investigated.

Under steady state conditions the balance of carbon dioxide or oxygen gas can be formulated as

$$D(d^2C/dz^2) - q = 0$$

where C is the concentration of the gas in g cm^{-3}, z is the vertical co-ordinate, D is the molecular diffusion coefficient in cm² s^{-1} and q is the source strength, positive for carbon dioxide and negative for oxygen, expressed in g cm^{-3} s^{-1}. For a constant q the solution to the differential equation is

$$C = a + bz + [q/(2\,D)]z^2$$

With the boundary condition $C = C_0$ at $z = 0$ the constant a becomes C_0 which is the atmospheric concentration. At the surface the balance of flux to total production rate $Q = qz_q$ (z_q being the root depth) requires that

$$-D(dC/dz) = -Q \quad \text{for } z = 0$$

the flux being negative because of its direction. With this $b = Q/D$ is obtained. Then the complete equation reads

$$C = C_0 + (Q/D)z + [Q/(2Dz_1)]z^2$$

for $0 < z < z_1$ and

$$C = C_0 + 1.5(Q/D)z_1 \quad \text{for } z > z_1$$

The diffusion coefficient D for carbon dioxide in air is close to 0.14 cm² s^{-1} (Albertsen, 1977). In porous media the effective diffusion distance is greater than the actual distance of flux. As a consequence of this the apparent diffusion coefficient in the gaseous phase in porous media is about 56% of the diffusion coefficient in free air (Albertsen, 1977, p. 47). In the present case the steady state situation is considered in which flux by diffusion is balanced by the source strength q. Therefore one must also consider the effect of porosity on the effective cross-section. This is the same as the porosity of the gas phase. If this is p, the effective diffusion coefficient is Dp where D is the diffusion coefficient in soil air, namely $0.14 \times 0.56 = 0.08$ cm² s^{-1}. In a soil with a gaseous pore space of 40% the effective diffusion coefficient is about 0.03 cm² s^{-1}. Since the yearly production of carbon dioxide was set to 1467 g m^{-1}, $Q = 4.66 \times 10^{-9}$. For a root depth of 50 cm one obtains

$$C_1 - C_0 = 1.166 \times 10^{-5} \text{ g cm}^{-3}$$

for the excess concentration of carbon dioxide below the root zone. Converted into pressure it becomes 0.0059 atm, i.e. the actual partial pressure would be 0.0062 atm or twenty times the atmospheric partial pressure of carbon dioxide. If the porosity of the gaseous phase is only 20% then the partial pressure of carbon dioxide in the soil below the root zone becomes 0.0121 atm. Increasing the root depth to 100 cm doubles the partial pressure below the root zone.

As for oxygen, the deficit partial pressure of oxygen gas will be numerically the same as the excess partial pressure of carbon dioxide, since diffusion coefficients are about the same and the same molar quantities are involved. Compared to the partial pressure of oxygen gas in the atmosphere, 0.2 atm, a deficit partial pressure of 0.01 atm is only 5% of the atmospheric pressure and would be imperceptible from a chemical point of view. In extreme cases – deep root zones and low porosities – the deficit partial pressure of oxygen gas could become large enough to cause reducing conditions below the root zone. Such conditions are probably rare.

Under water-logged conditions the gaseous porosity is zero. The diffusion coefficients in water solution are very small, of the order 1 cm^2 day^{-1}, thus about 1000 times smaller than in the gas phase. Consequently, a saturated root zone can be regarded as cut off from any oxygen gas supply. The organic matter decomposition into carbon dioxide and water will, however, continue with the aid of any constituents that can be reduced, ferric iron, manganese dioxide, nitrate ions and sulphate ions, since the energy release in the process is great anyway. Temporary saturations will, however, not make much impact since oxidizing conditions are soon restored but more prolonged situations will make a strong impact.

Albertsen (1977) studied carbon dioxide profiles in soil air in a few typical soils at different times of the year. According to the steady state theory given here the increase in partial pressure with depth should follow a second order polynomial up to the point where the root zone ends, after which it will remain constant. By and large, Albertsen's results confirm this. The seasonal variation in source strength, q, will cause some deviations from this rule in cultivated land while forest soils seem to keep to the 'rule'. The partial pressures observed by Albertsen, apparently below the root zone, vary from 0.001–0.004 atm in a coniferous forest soil, 0.001–0.006 atm in a deciduous forest soil, 0.003–0.015 atm in a grassland soil and 0.004–0.04 atm for cultivated land. The variation is, hence, considerable during the year but the range is well within that expected from the rate of turnover of organic matter. The partial carbon dioxide pressures in groundwater samples worked out from chemical analyses are also within this range.

The root zone is thus of importance for creating elevated partial pressures of carbon dioxide which set the weathering conditions in the intermediate zone. This zone will, in practice, be an open system with respect to carbon dioxide.

3.2.4 *Chemical weathering in the root zone*

It was pointed out in Section 3.2.1 that the root zone frequently shows signs of strong weathering in the past, the best evidence being the strongly leached part, the A horizon. Even in young podsolic soils the predominant mineral in the A horizon is quartz. The strong weathering which has taken place is no doubt due to the presence of organic matter forming complexes with iron and aluminium thus enhancing the rate of hydrolysis of primary minerals by removing the coating which otherwise seems to make them resistant. A great deal of material leached out of the A horizon is found in the B horizon, particularly iron and aluminium oxides, mixed with organic matter which may have facilitated the transfer downwards. Tamm (1940) investigated some podsol profiles in north Sweden. These profiles were developed during the last 9000 years, i.e. after the retreat of the inland ice. Table 3.7 shows the total loss of common constituents from the A horizon, worked out from analyses of the soil in the A horizon and in the parent material below the root zone. The amounts are impressive – nearly 2 kg m^{-2} of iron has been released. However, when converted into mean

Table 3.7 Weathering of primary minerals in the leached horizon of a podsol in the North of Sweden. The yearly rate was based on an age of 9000 years, the postglacial period (data from Tamm (1940))

Component	Total loss (g m^{-2})	Yearly average (mg m^{-2} year^{-1})
Aluminium	1500	170
Iron	1900	210
Magnesium	420	46
Calcium	270	30
Sodium	300	34
Potassium	350	30

yearly rates of weathering they are relatively small even when compared to deposition from the atmosphere. Since it is likely that the initial rate of weathering was much greater than that later on, the present rates of weathering in the root zone are probably small. Chemical weathering in the root zone does not contribute much to the concentration of dissolved components in soil water.

3.2.5 *Effects of changing biological activity in the soil*

It was pointed out earlier that the organic matter in the root zone is largely responsible for the ion exchange capacity of soils at least in groundwater recharge areas. Changes in organic matter content can therefore have strong effects on the chemical composition of soil water. A decrease will lead to a release of cations which will appear with the bicarbonate ion. An example may illustrate this point. The exchangeable calcium in the root zone of a normal podsol on glacial till in Sweden may amount to 50 g m^{-2}, which is 100 times the transport of calcium by percolation. If the organic matter content is reduced by 50% during a ten-year period, for example by clear-cutting a forest, the percolation rate of calcium would increase by a factor of six during this time. An increase would also occur for the other ions as well and for nitrate and sulphate through release of organically bound nitrogen and sulphur. The storage of these in organic matter is also large compared to the annual flow out of the root zone. Even slow changes in organic matter content could cause noticeable fluctuations in concentrations in percolating water.

3.2.6 *Effects of water-logging*

Water-logging in recharge areas may occur as a consequence of climatic changes or changes in land-use practices. High groundwater tables implies a flat topography combined with soils and rocks of low hydraulic conductivities. Changes in water balance due to increased precipitation or decreased evapotranspiration will cause rising groundwater levels and in places, particularly close to water divides, the groundwater surface may extend into the root zone.

When water-logging occurs the gas exchange between the root zone and the atmosphere is stopped or at least grossly inadequate for a normal turnover of organic matter, the reason being the small diffusion coefficient for carbon dioxide and oxygen in water as mentioned before.

Other means have to be used for oxidation. Consider Fig. 2.7 which refers to a system of oxygen gas, iron and water. The stability regions for different components are delineated in the usual manner. In this diagram the equilibrium between organic matter and carbon dioxide would be found at the top. Actually the stability area of glucose for example, extends well above the line for the stability of water with respect to decomposition into hydrogen and oxygen. When oxygen gas is present the system is represented by the bottom part of the diagram. When the oxygen gas is exhausted by the organic material present, the system moves upwards in the diagram. Any nitrate present would be converted into nitrite in the lower part of the diagram. In the upper half of the diagram the ferric oxides are converted into ferrous iron. The process can be written

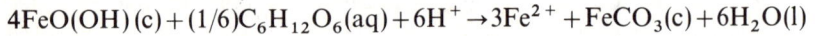

$$4FeO(OH)(c) + (1/6)C_6H_{12}O_6(aq) + 6H^+ \rightarrow 3Fe^{2+} + FeCO_3(c) + 6H_2O(l)$$

when the pH is high enough to form ferrous carbonate, siderite. It is, however seen that only a quarter of the iron reduced forms carbonate, the rest must be in solution. From standard free energy data it can be shown that the pH will be as high as 8.4 before equilibrium is reached. The buffering capacity of soils will most likely never allow such a high pH hence all ferric iron will be reduced to ferrous iron as long as there is organic matter present. When no siderite is present, it is postulated that bicarbonate ions will form. The system is, however closed so that the bicarbonate will make up only one quarter of released ferrous iron. Under these conditions the equilibrium pH would be somewhat lower, just above 8, but still high enough to be unrealistic in a leached soil. In an alkaline soil it would be feasible for siderite, or even ferrous hydroxide to be formed. It should be noted that the equilibria discussed are really on the upper border of the stability region for water.

Sulphate can also act as an oxygen source for organic matter. In a water-logged soil it would form pyrite, FeS_2. The amount of sulphate available for this is usually very small compared to the iron present in the B horizon. All would therefore be fixed as pyrite after which the reduction proceeds with ferric oxides as described.

Concluding this subsection it can be stated that the very high concentrations of ferrous iron sometimes observed in groundwater most likely originate from B horizons of soils developed under less humid conditions than the present. This implies that present peatlands in temperate climates were once at least partly well drained groundwater recharge areas. During the so-called climatic optimum some 3000 years back the peatland area was presumably much smaller than at present.

Water-logging will favour accumulation of organic matter and hence will also lock up cations in exchange positions. Also nitrogen and sulphur will accumulate as organic compounds. A slow but persistent increase in storage of various elements is thus expected. Severe water-logging will, of course, induce peat formation.

3.3 Processes in the intermediate zone

3.3.1 Definitions and hydrological concepts

The intermediate zone is the part of the ground below the root zone which has a gaseous phase in contact with the atmosphere through the gaseous phase of the root zone. Water in the intermediate zone occupies the smallest pore spaces. It moves downward at a rate set by the hydraulic conductivity. This depends on the water content. When water content is low the hydraulic conductivity is so small that the downward flow of water becomes imperceptible on a time scale of days. The water content in the intermediate zone under such condition is referred to as 'field capacity'. The term field capacity should not be used without specifying the time scale involved.

Flow of water in the intermediate zone, percolation, takes place with little longitudinal dispersion as evidenced by the horizontal stratification of various dissolved components observed. An example of the distribution of chloride in solution in profiles from the Botswana part of the Kalahari desert is shown in Fig. 3.5 (Foster *et al.*, 1982). In its simplest version the flow can be regarded as so-called piston flow, i.e. without any longitudinal dispersion. When required, longitudinal dispersion can be considered through a partial differential equation of the following kind

$$dC/dt + w(dC/dz) = D(d^2C/dz^2) + q(z)$$

which is a mass balance equation. In this equation w is the vertical flow velocity of water, C the concentration of a substance, z the vertical co-ordinate, D a diffusion coefficient and $q(z)$ the source strength adding or subtracting the substance. If $q(z) = 0$ and $D = 0$ the solutions become simple. When C is initially zero but is kept at $C = C_0$ at $z = 0$ from $t = 0$ then the solution is

$$C = C_0 \quad 0 < z > wt$$
$$C = 0 \quad z > wt$$

thus a true piston flow illustrated in Fig. 3.6 by the full drawn line. When dispersion has to be considered the concentration with depth will follow a curve like the dashed one.

The travel time for water through the intermediate zone depends on the water content of the zone and the average flow rate. Using mm of water as a unit the water content may be as much as 200 mm m^{-1} depth. For a yearly groundwater recharge of 200 mm the travel time for water becomes 1 year m^{-1}. For a 10 m deep intermediate zone under the same conditions

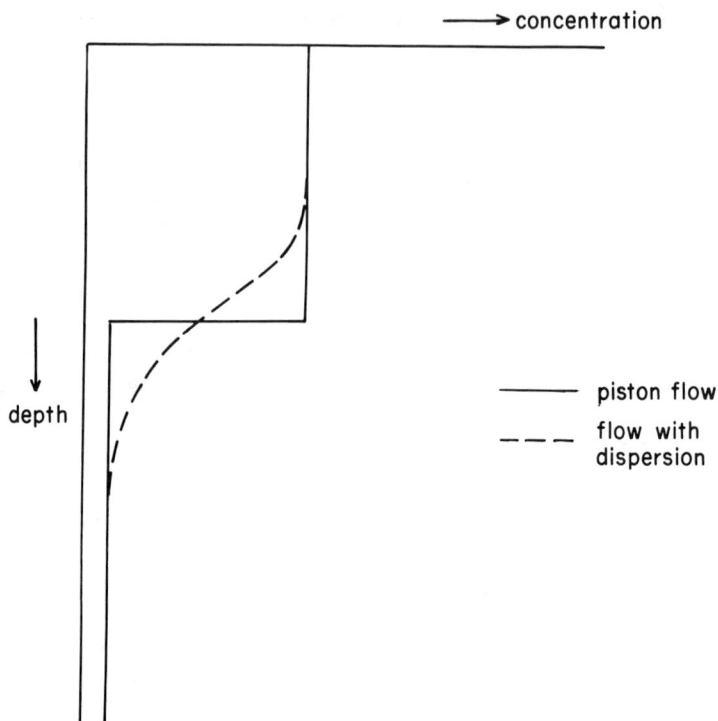

Fig. 3.6 Concentration profiles for piston flow and flow with longitudinal dispersion expected in the unsaturated zone, in which a tracer is added at a constant rate from a fixed time.

it will take ten years for the water leaving the root zone to reach the groundwater surface.

Consider now an intermediate zone which has a water content close to 'field capacity'. At a certain moment a quantity of water is added from the root zone, being an excess of what the root zone can keep for the moment. This addition increases the water content in the uppermost part of the zone which also increases the hydraulic conductivity. As a result the previous water is pushed downward and, because of flux convergence, is compressed, i.e. increases the water content in this part. However, this causes a continued 'compaction' of water in the intermediate zone, a compaction that moves down at a speed which depends on the hydraulic conductivity in the 'compacted' state. It soon reaches the bottom of the intermediate zone, i.e. the groundwater surface which receives an amount of water equal to that added from the root zone. What has happened is actually a displacement of the whole water column because of the addition at the top.

Since the water content in the intermediate zone can vary, the displacement will occur as a travelling pulse. The time it takes for the displacement to reach the groundwater surface is called the response time of the water in the intermediate zone. This response time is much shorter than the transit time. For a 10 m deep intermediate zone the response time may be a few weeks only, while the transit time is perhaps ten years.

Hydrodynamically, the intermediate zone is a fairly simple and predictable system with a relatively rapid response to water additions. Chemically it is a considerable reservoir of water in very intimate contact with soil air and, because of this, with the atmosphere. Chemically it is an open system with respect to oxygen and carbon dioxide.

3.3.2 *Gas exchange, intermediate zone–root zone*

The gas exchange rate is set by the diffusion coefficients for air constituents. The partial pressures of oxygen and carbon dioxide in the intermediate zone will tend to follow those in the root zone. Seasonal variations in the intensity of organic decomposition in the root zone will be reflected to some extent in the intermediate zone, though only to the extent that gaseous diffusion is able to 'fill up' or 'deplete' storage so as to equalize it with the root zone. In the deepest part of the intermediate zone seasonal fluctuations will be small, lagging in phase while in the upper part the fluctuations will follow those in the root zone. Some carbon dioxide will be consumed during chemical weathering but quantities involved are extremely small when compared to the transport capacity for carbon dioxide in the intermediate zone by molecular diffusion. The same is true for oxygen.

3.3.3 *Chemical reactions in the intermediate zone*

The chemical reactions in the intermediate zone can be grouped into three categories:

(a) Dissolution of calcite
(b) Hydrolysis of primary minerals
(c) Synthesis of clay minerals

(b) and (c) being also what have been termed 'mineral transformations'. When calcite is present in the intermediate zone this will in general react rapidly with carbon dioxide, thus changing the concentration of the water drastically. Concurrently, the very much slower processes of mineral transformations (hydrolysis of primary minerals and formation of clay minerals) will take place but may have much less impact on the salt concentration than calcite dissolution.

Calcite dissolution

The reaction which takes place can be written

$$CaCO_3(s) + H_2O(l) + CO_2(g) \leftrightarrow Ca^{2+} + 2HCO_3^-$$

Calculating the thermodynamic equilibrium constant at, say, $+10°C$ is done by the following procedure (ignoring ion pairs)

$$\Delta G_r° = -132.18 - 2 \times 140.31 + 269.89 + 56.69 + 94.26$$
$$= 8.04$$
$$pK = 5.89$$
$$\Delta H_r° = -129.8 - 2 \times 165.2 + 289.0 + 68.32 + 94.1$$
$$= -8.78$$
$$pK_{10} = 5.89 - (8.78 \times 15)/(1.3643 \times 283.15)$$
$$= 5.89 - 0.34 = 5.55$$

Hence

$$pCa + 2pHCO_3^- = 5.55 + pPCO_2$$

This is strictly valid only for activities of calcium and bicarbonate. However, the reaction produces twice as many moles per litre of bicarbonate as of calcium. In the same notation as in the equation above this condition reads

$$pCa = pHCO_3 + 0.3$$

Introducing pf_1 and pf_2 as the negative logarithms of the activity coefficients of bicarbonate and calcium respectively one can write

$$3pHCO_3 = 5.25 + pPCO_2 - 2pf_1 - pf_2$$

which shows that the bicarbonate concentration, as well as the calcium concentration, will vary proportionately with the cubic root of the carbon dioxide partial pressure. Neglecting for the moment the activity coefficients, assuming carbon dioxide in the intermediate zone to be 0.001 atm (30 times the partial pressure of carbon dioxide in the atmosphere) the bicarbonate concentration becomes 1.78 mmol l^{-1} or 108 mg l^{-1}. Then there must be 0.89 mmol l^{-1} of calcium or 36 mg l^{-1}. If the carbon dioxide partial pressure is increased to 0.008 atm the concentrations of calcium and bicarbonate will double.

Considering the activity coefficients one can use the Debye–Hückel equations in the form

$$f = 0.5\sqrt{I}/(1 + 1.3\sqrt{I})$$
$$f = 2\sqrt{I}/(1 + 1.3\sqrt{I})$$

and for an ionic strength of 0.01, corresponding roughly to an electric conductivity of 70 mS m^{-1}, $pf_1 = 0.044$ and $pf_2 = 0.176$. Hence

$$-2pf_1 - pf_2 = -0.22$$

Then at a partial pressure of carbon dioxide equal to 0.001 atm there will be 2.11 mmole l^{-1} of bicarbonate, i.e. 129 mg l^{-1} and 42 mg l^{-1} of calcium.

The pH is obtained from the reaction

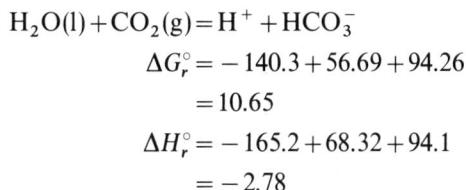

$$H_2O(l) + CO_2(g) = H^+ + HCO_3^-$$
$$\Delta G_r^\circ = -140.3 + 56.69 + 94.26$$
$$= 10.65$$
$$\Delta H_r^\circ = -165.2 + 68.32 + 94.1$$
$$= -2.78$$

Hence, since $pK = 7.81$

$$pK_{10} = 7.81 - (2.78 \times 15)/(1.3643 \times 283.15)$$
$$= 7.70$$

Then

$$pH + pHCO_3 = 7.70 + pPCO_2$$

For a carbon dioxide pressure of 0.001 atm it was found that the bicarbonate concentration was 2.11 mmol l^{-1} and thus $pHCO_3 = 2.67 + 0.04 = 2.71$. Then

$$pH = 7.7 + 3 - 2.71 = 7.99$$

and at a partial pressure of 0.008 atm

$$pH = 7.7 + 2.1 - 2.41 = 7.39$$

Actually, the hydrogen ion activity will be proportional to the square of the cube root of the partial pressure of carbon dioxide.

The reaction between carbon dioxide and calcite is comparatively rapid. If calcite is present in the intermediate zone the sinking water will soon get saturated with calcite and the calcium concentration will be set by the partial pressure of carbon dioxide. Also the pH will be determined by this, provided there was no acidity or alkalinity in the water that entered the intermediate zone.

The reaction scheme used here excludes the carbonate ion. This is permissible as long as the pH does not exceed 8.5. The minus logarithm of the second dissociation coefficient of carbon dioxide exceeds 10 so that at pH = 8.5 the ratio carbonate to bicarbonate is less than 0.03. The highest pH would occur for the carbon dioxide partial pressure of the atmosphere.

Then the bicarbonate concentration would be 1.41 mmol l^{-1} and pH $= 8.37$.

Reaction rates involving other minerals

The chemical reactions which are of interest here can be classified as ionic exchange, adsorption, hydrolysis of primary minerals and formation of clay minerals. Ionic exchange is only important as a buffer mechanism during transient states as pointed out before. Evidence of ionic exchange detected through chemical balance studies can reveal changes in organic matter content or effects of acid precipitation. The intermediate zone is the first site after the root zone to take the impact of such changes. Considering that the piston flow type of water motion (with some dispersion) is well documented in the intermediate zone, the disturbance caused by any of the changes mentioned would travel down the profile as fronts, with speeds that depend on the exchange capacity and adsorption capacity.

Turning attention to 'weathering' processes and their effect on primary minerals, this is basically a hydrolysis process, the minerals being formed under quite different moisture conditions. Hydrolysis releases silica from the primary minerals and since silicic acid is extremely weak there will be a rise in the pH under any conditions. Consider the reaction between potash feldspar, microcline, and water

$$KAlSi_3O_8(c) + 2H_2O(l) \leftrightarrow K^+ + OH^- + Al(OH)_3(c) + 3SiO_2(am)$$

illustrating the production of hydroxyl ions. In the presence of carbon dioxide one would get

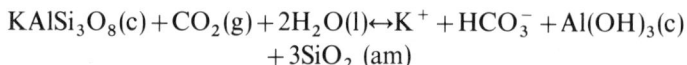

$$KAlSi_3O_8(c) + CO_2(g) + 2H_2O(l) \leftrightarrow K^+ + HCO_3^- + Al(OH)_3(c) + 3SiO_2 \ (am)$$

and the pH would be set by the bicarbonate ions produced and the partial pressure of carbon dioxide as discussed earlier. As the reaction is written, aluminium hydroxide is formed and the study by Busenberg and Clemency (1976) confirms that during the initial phase the ionic product $(Al^{3+})(OH^{-3})$ is that of amorphous aluminium hydroxide. The remarkable thing about the reaction is that it is quite fast in the very beginning but rapidly slows down during the first few days. After that the rate of dissolution is inversely proportional to the square root of time. The explanation offered is that a coating of aluminium hydroxide and silica builds up, slowing down the diffusional transport to and from the surface of the mineral. In the final stage, after a few hundred hours, the rate of reaction becomes constant and comparatively small. Busenberg and Clemency also obtained some indication from the Al:Si ratio of the coatings, that in the case of potash feldspars this ratio was 1:2 which can be interpreted to mean that Si–Al–Si clay layers were formed, probably as a beginning of illite formation. For feldspars poorer in silica anorthite, for example, the Al:Si ratio became about 1:1 indicating kaolinite formation. It is doubtful that under the conditions described the partial pressure of carbon dioxide is setting the rate of transformation of the primary minerals. Inside the 'coatings' the conditions are probably close to equilibrium with micro-crystalline phases of the appropriate clay minerals.

The rate constants obtained experimentally are expressed in $mol\ cm^{-2}\ s^{-1}$ so that the exposed surface area also determines the amounts released. The rate constants are small, their logarithm being around -16, considered by Busenberg and Clemency to represent a pH of about 5 since they used dissolved carbon dioxide at a pressure of about 1 atm during the experiments. At a partial pressure of about 0.01 atm one would expect perhaps a ten times lower reaction rate since, according to the reaction written. The equation is

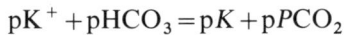

$$pK^+ + pHCO_3 = pK + pPCO_2$$

at silica saturation. The concentration of K inside the coating is, as can be seen, the same as the concentration of bicarbonate.

In order to estimate the effect of weathering of primary minerals on the chemistry of water in the intermediate zone, one needs to know both the specific mineral surface area, the water content and the flow rate through the layer. If this is of the order 1 year per metre depth at a water content of $200\ mm\ m^{-1}$ depth and a specific mineral surface area (of a feldspar) of $100\ m^2\ m^{-3}$, the rate of release becomes about 30 micromoles per litre per year. For sodium this means 0.7 mg per litre per year. At an intermediate zone depth of 10 m the sodium concentration at the bottom would become $7\ mg\ l^{-1}$. A surface area of $100\ m^2\ m^{-3}$ is perhaps small though not unreasonable. In fissured rocks the specific surface area is probably much less.

Reaction rates have also been studied for other minerals. Luce *et al.* (1972) investigated some magnesium silicates. They also found constant dissolution rates for very long periods of time but the order of magnitude of the rate constant was about the same as that for feldspars.

The release of cations by silicate weathering will influence the pH in a predictable manner. Consider the previous reaction with equilibrium between bicarbonate and carbon dioxide as determined by the reaction

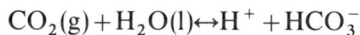

$$CO_2(g) + H_2O(l) \leftrightarrow H^+ + HCO_3^-$$

Using standard free energy data one gets

$$\Delta G_r^\circ = -140.31 + 94.26 + 56.69 = 10.64$$

Hence pK becomes 7.78 and the equilibrium equation

$$pH + pHCO_3 = 7.78 + pPCO_2$$

For a bicarbonate concentration of 2 mmol l^{-1}, pHCO$_3$ = 2.7. With a partial pressure of carbon dioxide of 0.01 atm, pPCO$_2$ = 2, pH becomes 7.08. With a bicarbonate concentration of 0.1 mmol l^{-1}, a fairly common level in non-calcareous rock areas and with the partial pressure of carbon dioxide of 0.001 atm, the pH becomes $7.78 - 4 + 3 = 6.78$. The pH of water in the unsaturated zone is thus determined by the degree of weathering which has taken place and by the partial pressure of carbon dioxide in the soil air. The bicarbonate concentration, mostly equivalent to alkalinity, matches the cations released during weathering provided no strong acid is present.

3.3.4   *Mineralogical zonation*

It was pointed out, when discussing the root zone, that the combination of chemical weathering and vertical transport creates a zonation of the soil profile with respect to products of weathering. This is particularly evident in podsolic soils where the minerological part of the A horizon sometimes consists of almost pure quartz, the aluminium and iron hydroxides being dissolved and removed into the adjacent B horizon. Under the climatic conditions where podsolic soils are formed, quartz is definitely the most resistant mineral while iron and aluminium hydroxides are next on the list of stability. Knowing the climatic conditions and the rock material it is possible to construct a list of the various minerals to be found, ranked according to their stability under prevailing conditions. Such a list could, for instance, be quartz, iron and aluminium hydroxides, kaolinite and montmorillonite. Formation of gibbsite, i.e. aluminium hydroxide, is favoured when the rate of groundwater recharge is so high as to preclude saturation by amorphous silica. Under less humid conditions the silica concentrations may become high enough to make kaolinite more stable than gibbsite. However, even under humid conditions the silica concentration at deeper levels may become large enough to preclude formation of gibbsite. A sequence of quartz, gibbsite and kaolinite is therefore possible in the direction of water flow. Deeper down illite minerals may form, since they require still more silica.

3.4   Processes in the water saturated zone (groundwater)

This zone is characterized by full water saturation; even the so-called capillary fringe is saturated. The border between intermediate and saturated zones may in many cases be irregular because the capillary properties of the material may vary. One can, however, add the condition that the major diffusional transport in the saturated zone has to take place in the liquid phase even when averaged horizontally over a distance of, say, 0.1 m.

With this description and criteria the saturated zone will be a closed system with respect to gases, in contrast to the intermediate zone. The immediate conclusion of this is that carbon dioxide and oxygen gas present in the saturated zone originate from that dissolved at the moment they entered the saturated zone. There is, thus, no replacement for oxygen consumed nor any escape for carbon dioxide produced inside the saturated zone or emanating from below. The internal production of carbon dioxide deserves special attention and will be discussed in some detail. Since such production usually involves oxidation of organic matter also the oxygen and oxidation/reduction processes are linked to carbon dioxide production and need to be considered in various contexts.

3.4.1 *The solid matrix of the saturated zone*

The medium of saturation can consist either of porous material like sandstone or of fissures in rocks that have been subjected to brittle deformation. In old limestone areas a third type, karst, is found consisting of solution channels developed along fractures in the limestone. A combination between pores and fractures can be found in old sandstones which are partly metamorphosed and between fractures and solution channels in limestones.

The flow pattern in pores and fractures is in general considered to be laminar, i.e. completely stationary in time unlike turbulent flow probable in solution channels. In this case flow patterns have to be described in a statistical sense as time averages.

Pores, fractures and solution channels are in a wide sense randomly distributed. This is particularly true for pores in homogeneous rocks. Fractures, being a result of brittle deformation, are on a small scale also randomly distributed in space but on a larger scale fracture zones can be identified.

Not all fractures can transport water. Some formed during shear stress may be closed in a sense that there is no exit for water and, sometimes, not even an entrance. Others, developed during tensional stress are wide with ample connections along fracture zones. Vertical fractures of that type may reach considerable depths, the deepest being often filled with basic rock material. Near-horizontal fractures are found fairly close to the rock surface and are considered to be due to release of vertical pressure when the rock emerged from great depths.

Solution channels, although almost exclusively found in limestones, probably developed along fractures or on contact surfaces to other rock types. There are spectacular ones forming caves and long tunnels which may be more or less accessible.

The water content (at saturation) of various rocks varies considerably. In porous unconsolidated material like gravel, sand and silt it may amount to 30% by volume. In sedimentary rocks like sandstones it may reach 20%. In fractured rocks it is often less than 1% if averaged over large volumes. In some rocks like strongly fractured and fissured limestones the pore volume may be considerable.

The hydraulic conductivity of porous rocks also varies considerably depending primarily on pore size. In coarse, weakly consolidated material it may exceed 10^{-6} m s^{-1} while in siltstone it is very much less and in clays and shales perhaps below 10^{-9} m s^{-1}. Fractured rocks can have exceptionally high hydraulic conductivities in fracture zones while more solid rock masses have conductivities of less than 10^{-9} m s^{-1}. Fracture zones and solution channels can often be considered as natural drainage channels. The very deep open fractures are of particular importance in this case.

The flow pattern in the saturated zone is complex, seen in a small scale. However, on a large scale they will appear quite predictable and are frequently simulated by mathematical models. The scale to choose in such studies depends on the water-bearing structures.

Flow in porous and fractured media are subject to dispersion, i.e. mixing of dissolved substances during the transport, particularly in the direction of flow. This is then called longitudinal dispersion. The dispersion prevents a sharp zonation of chemical characteristics along the flow direction of water. The dispersion can be modelled by applying the diffusion equation of the type

$$\partial C/\partial t + u(\partial C/\partial x) = \partial/\partial x(D\ \partial C/\partial x) + q$$

C being concentration, D being a diffusion coefficient, u the average velocity in the x-direction and q the source (or sink) strength.

There is at present no way to calculate D from the characteristics of the aquifer. Also, D depends on the scale under consideration, increasing with increasing scale. D is related to the structure of pores and fractures. If L is the distance required to attain near average flow velocity then L is the scale of dispersion, and D should be proportional to Lu. L would also depend on the lateral scale of flow unless the medium is very homogeneous in a large scale sense.

64

3.4.2 *Gas exchange between the intermediate zone and the saturated zone*

The gas exchange between the two media is determined by the molecular diffusion rate in water and the solubility of the gas in water. The net flow of gas into the saturated zone has a minimum which equals the concentration of gas in the water times the rate of groundwater recharge. Any additional gas inflow to the saturated zone requires a negative concentration gradient, i.e. concentration decreases in the direction of groundwater flow.

It is instructive to compare concentrations of gases like carbon dioxide and oxygen in water and in the gas phase. In the gas phase, 1 mole of a gas at 0°C and 1 atm pressure occupies a volume of 22.4 litres. The concentration in the gas phase expressed as $mol\,l^{-1}$ is consequently 0.049 or 49 $mmol\,l^{-1}$. In water at 1 atm of carbon dioxide pressure 77 $mmol\,l^{-1}$ dissolves at 0°C while oxygen gas in water at 0°C and 1 atm is 2.18 $mmol\,l^{-1}$ (see Table 2.2). Carbon dioxide is thus about 30 times more soluble than oxygen gas in water. Consider a carbon dioxide partial pressure in the intermediate zone of 0.01 atm and a partial pressure of oxygen gas of 0.2 atm, the atmospheric value. Then there will be 0.77 $mmol\,l^{-1}$ of carbon dioxide and 0.44 $mmol\,l^{-1}$ of oxygen gas in the water which is recharged provided the temperature is at 0°C. At 10°C one would have 0.54 $mmol\,l^{-1}$ of carbon dioxide and 0.34 $mmol\,l^{-1}$ of oxygen gas, other conditions being the same. Considering the high partial pressure of carbon dioxide in the root zone and in the intermediate zone this leads to comparable concentrations of dissolved carbon dioxide and oxygen gas in the water that is recharging the groundwater in the saturated zone.

For a recharge rate of 100 mm yr^{-1}, 100 litres of water is added per metre which means about 50 mmol per square metre per year of carbon dioxide considering the situation just discussed. The molecular diffusion coefficient for carbon dioxide in water is around 1 $cm^2\,day^{-1}$ equal to 0.0365 $m^2\,yr^{-1}$. The gradient required to give a transport by diffusion of 50 $mmol\,m^{-2}\,yr^{-1}$ then becomes $50/0.0365 = 137\ mmol\,m^{-3}\ m^{-1}$. Comparing this to the equilibrium concentration 0.5 mmol $l^{-1} = 500\ mmol\,m^{-3}$ it is seen that the concentration must decrease by 0.137 $mmol\,l^{-1}$ for every metre. Thus after $0.5/0.137 = 3.65$ m the carbon dioxide concentrations should be zero in order to enable the required diffusional transport. Now, the calculated diffusional transport is certainly too high since the space available for diffusional transport is at best only 20%. The gradient therefore needs to be five times greater than was worked out before, so the zero concentration distance becomes only 0.73 m. A really strong sink for carbon dioxide is thus required close to the groundwater surface in order to maintain the diffusional transport required in this example. In the absence of such a sink it is evident that diffusional transport of carbon dioxide and oxygen into the saturated zone is very small. The major means of transport for gases into the saturated zone is by recharge water from the intermediate zone. The saturated zone can be considered for all practical reasons, a closed system with respect to dissolved gases unlike the root zone and the intermediate zone.

3.4.3 Chemical reactions in the saturated zone

There is no principle difference between the saturated and intermediate zones with respect to chemical reactions since no new components need to be considered. However, some modifications in the formulation of reactions may be useful. Since partial pressures of gases are no longer maintained by a gaseous phase, reactions should contain only crystalline species and species in solution.

Reactions involving calcite
The basic reaction is then

$$CaCO_3(c) + H_2O(l) + CO_2(aq) \leftrightarrow Ca^{2+} + 2HCO_3^-$$

The solubility of calcium carbonate should also be taken into account

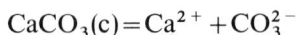

$$CaCO_3(c) = Ca^{2+} + CO_3^{2-}$$

and the equilibrium between bicarbonate and carbonate ions

$$HCO_3^- = H^+ + CO_3^{2-}$$

in case the final pH becomes high.

In addition to these equilibria one must consider the mass balance and the charge balance. They depend on the initial concentrations of inorganic carbon species, defined by

$$\sum[CO_2] = [CO_2] + [HCO_3^-] + [CO_3^{2-}]$$
$$A = [HCO_3^-] + 2[CO_3^{2-}] + [OH^-] - [H^+]$$

$\sum[CO_2]_0$ and A_0 can be taken to represent initial total inorganic carbon and alkalinity respectively. The mass and charge balances can then be formulated

$$\sum[CO_2]_0 + \Delta[Ca^{2+}] = \sum[CO_2]$$
$$A_0 + 2\Delta[Ca^{2+}] + \Delta[H^+] = A + \Delta[OH^-]$$

where the Δ symbols indicate the changes which have taken place. If calcite solution equilibrium was established in the intermediate zone under constant partial pressure of carbon dioxide nothing additional will happen. Consequently, $\Delta[Ca^{2+}]$ will be zero as well as $\Delta[H^+]$ and $\Delta[OH^-]$. The carbon dioxide concentrations will remain the same as before. There may be other changes later which cause adjustments in the calcite equilibrium but this will be discussed in the next subsection.

The other extreme case worth considering is when no dissolution of any kind has taken place in the intermediate zone but there is calcite present in the saturated zone. In this case, $\sum[CO_2]_0 = [CO_2]_0$, the initial carbon dioxide concentration. A_0 is supposed to be zero since the very small concentration of HCO_3^- is balanced by the concentration of H^+.

The complete set of equations governing the equilibrium then becomes

$$pCa + 2pHCO_3 = pK' + pCO_2 \tag{3.1}$$
$$pCa + pCO_3 = pK_s' \tag{3.2}$$
$$pH + pCO_3 = pK_2' + pHCO_3 \tag{3.3}$$
$$[Ca^{2+}] + [CO_2]_0 = [CO_2] + [HCO_3^-] + [CO_3^{2-}] \tag{3.4}$$
$$2[Ca^{2+}] + [H^+] = [HCO_3^-] + 2[CO_3^{2-}] + [OH^-] \tag{3.5}$$

The element symbols all refer to concentrations. The equilibrium coefficients are related to the thermodynamic constants and the activity coefficients by

$$pK' = pK - pf_2 - 2pf_1$$
$$pK_s' = pK_s - 2pf_2$$
$$pK_2' = pK_2 - pf_2 + pf_1$$

Using the Debye–Hückel approximations of activity coefficients in the form

$$pf_z = 0.5\ z^2\sqrt{I}/(1+\sqrt{I})$$

with I being ionic strength, one can write

$$pK' = pK - 6pf_1$$
$$pK_s' = pK_s - 8pf_1$$
$$pK_2 = pK_2 - 3pf_1$$
$$pf_1 = 0.5\sqrt{I}/(1+\sqrt{I})$$

The set of equations (3.1)–(3.5) can now be used to determine concentrations of calcium, dissolved carbon dioxide, bicarbonate, carbonate, and the pH. Since the pH will always be above 7, $[H^+]$ can be ignored.

Initially $[CO_2]_0$ is given by the temperature and the carbon dioxide partial pressure in the intermediate zone.

The set of equations can be solved by iteration. An example of computed concentrations of the various constituents is shown in Table 3.8. The temperature was set at 5°C and the ionic strength at 0.0005. Under these conditions one obtains $pK' = 4.09$, $pK'_s = 8.13$ and $pK'_2 = 10.44$.

The results are interesting since they show that the solubility of calcite becomes of importance when the dissolved carbon dioxide concentration is low, and also that the following reaction predominates

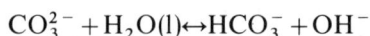

$$CO_3^{2-} + H_2O(l) \leftrightarrow HCO_3^- + OH^-$$

The result in these conditions is a groundwater with a fairly high pH and low calcium concentration, i.e. about 7 mg l^{-1} in the first case considered. In an open system with the corresponding partial pressure of carbon dioxide, in the same conditions, the calcium concentration would have been around 40 mg l^{-1}.

Variations in ionic strength have a rather small effect on equilibrium concentrations in the range 0.0001 to 0.01. Temperature influences the solubility of carbon dioxide and thus the initial carbon dioxide concentration. Table 3.9 gives some computed data at 5°C and 10°C and two levels of initial concentrations of carbon dioxide.

For equal initial concentrations of carbon dioxide the differences due to temperature are very small. However, when the initial carbon dioxide concentrations are set by the same partial pressures of carbon dioxide in the intermediate zone, then the differences with temperature become appreciable for high levels of the partial pressure. It is seen in the table that an increase in temperature of 5°C lowers the initial concentrations by almost 20%. At the low initial carbon dioxide level (corresponding to 0.001 atm partial pressure of carbon dioxide) the effect of the decreased carbon dioxide initial concentration is small anyway because the solubility under hydrolysis of calcite is dominating.

The more general case with more complicated initial conditions can be handled in the same way. The general case would consider initial concentrations of bicarbonate and carbonate as well as of calcium.

The effect of mineral transformations in the saturated zone on calcium
Mineral transformations or the weathering of primary minerals described in Chapter 2 are, as a rule, slow processes, the rates of reaction being governed by the physical properties of the coating formed on the surfaces of the minerals. As pointed out earlier, most primary minerals will react with water, resulting in an addition of silica, cations and hydroxyl ions to the solution. In so far as carbon dioxide is present it will be converted into

Table 3.8 Equilibrium concentrations of calcium, bicarbonate and carbonate and equilibrium partial pressure of carbon dioxide in a closed system for different initial carbon dioxide partial pressures. (There were no carbonate species initially)

Initial		At equilibrium				
$[CO_2]$ $(mmol\ l^{-1})$	PCO_2 (atm)	$[Ca^{2+}]$	$[HCO_3^-]$ $(mmol\ l^{-1})$	$[CO_3^{2-}]$	pH	PCO_2 (atm)
0.064	0.001	0.170	0.191	0.044	9.8	0.0000004
0.128	0.002	0.200	0.291	0.037	9.5	0.0000033
0.256	0.004	0.294	0.525	0.025	9.1	0.0000156
0.512	0.008	0.530	1.028	0.014	8.6	0.0001
1.024	0.016	1.032	2.048	0.007	8.0	0.00083
2.048	0.032	2.052	4.096	0.003	7.3	0.0066

Table 3.9 Effects of temperature on calcite equilibria in closed systems

Temperature (°C)	5	10	10	5	10	10
Initial [CO_2]	0.064	0.064	0.054	0.64	0.64	0.54
Equilibrium [Ca^{2+}]	0.170	0.167	0.164	0.653	0.653	0.556
Equilibrium [HCO_3^-]	0.191	0.189	0.175	1.282	1.282	1.083
Equilibrium [CO_3^{2-}]	0.044	0.042	0.043	0.011	0.011	0.013
Equilibrium pH	9.80	9.75	9.80	8.38	8.33	8.48

bicarbonate and carbonate by the hydroxyl ions at the same rate as the weathering proceeds. It is doubtful if dissolved carbon dioxide itself can speed up the transfer of ions through the mineral coating unless it can attack the coating such that the diffusion of ions through it increases. It may, as pointed out earlier, influence the equilibrium at the mineral surface in a predictable way but so far there is no experimental evidence of this.

The carbon dioxide will thus decrease with time. If the transit time of water between the recharge area and the discharge area is relatively short, dissolved carbon dioxide would still exist in solution. If the transit time is long all the carbon dioxide is likely to be completely converted into bicarbonate or carbonate, accompanied by common ions sodium, magnesium, potassium and calcium. There may be some internally produced carbon dioxide from reactions between dissolved oxygen and organic matter. However, even if all oxygen is consumed in this process it may at best double the usable storage of carbon dioxide as pointed out earlier.

When calcium is present in solution it may be affected by the decreasing carbon dioxide concentration. Hence, at some point the reaction

$$CaCO_3(c) + CO_2(aq) + H_2O(l) \leftrightarrow Ca^{2+} + 2HCO_3^-$$

will begin to run from right to left precipitating calcium carbonate and thereby releasing carbon dioxide, which is then used in the mineral transformations. A somewhat simpler way to look at this is by considering the reaction

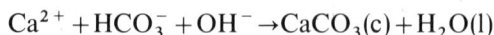

$$Ca^{2+} + HCO_3^- + OH^- \rightarrow CaCO_3(c) + H_2O(l)$$

which shows how hydroxyl ions are consumed through precipitation of calcium carbonate. There is plenty of evidence of calcite formation in fractures and fissures in igneous rocks, no doubt because of mineral transformations releasing hydroxyl ions. For a pure calcium bicarbonate solution half of the bicarbonate would be used up in this way precipitating all calcium as time passes, leaving a solution of potassium or, more probably, sodium bicarbonate. Figure 3.7 demonstrates schematically the effect of mineral transformation of a sodium rich rock on the calcium and sodium concentrations. In (a), full calcite saturation was established at the very beginning while albite weathering proceeds with time; in (b), some of the calcite was unsaturated initially; in (c), finally there was no initial calcite dissolution, just the slow weathering of a plagioclase rock releasing calcium and sodium. In this case, calcium increases with time to begin with but is precipitated gradually as the dissolved carbon dioxide is used up, leaving

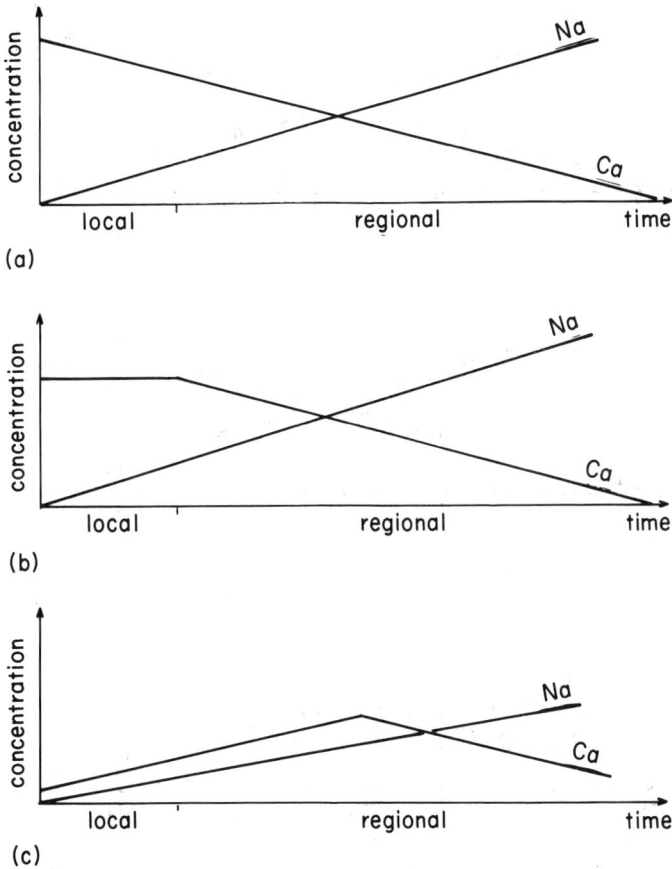

Fig. 3.7   The time development of calcium and sodium along the flow path when (a) calcium saturation was established at the start and albite weathering is taking place; (b) calcium unsaturation prevailed at the start but albite weathering is taking place; (c) calcite unsaturation prevailed at the start but plagioclase weathering is taking place.

sodium bicarbonate in solution. The time scale is indicated by the labels 'local', 'intermediate' and 'regional' referring to the type of groundwater transit times. Local groundwater flow will have a transit time for water in the order of ten years and will rarely show any evidence of calcite precipitation, since this time is likely to be too short for any more extensive mineralization through mineral transformation. The transit time of the intermediate groundwater flow system will be perhaps an order of magnitude greater but may still be too short to reach the critical state where calcite starts to precipitate. The regional flow, however, will most likely occupy the major groundwater volume and have transit times of 1000 years or more. Here the effect of mineral transformation would become evident forming sodium bicarbonate waters. Figure 3.8 shows schematically the border of calcite precipitation in a granite aquifer indicating the different flow systems.

The effects of mineral transformations on calcite precipitation just described are, of course, limited to such rocks that contain primary

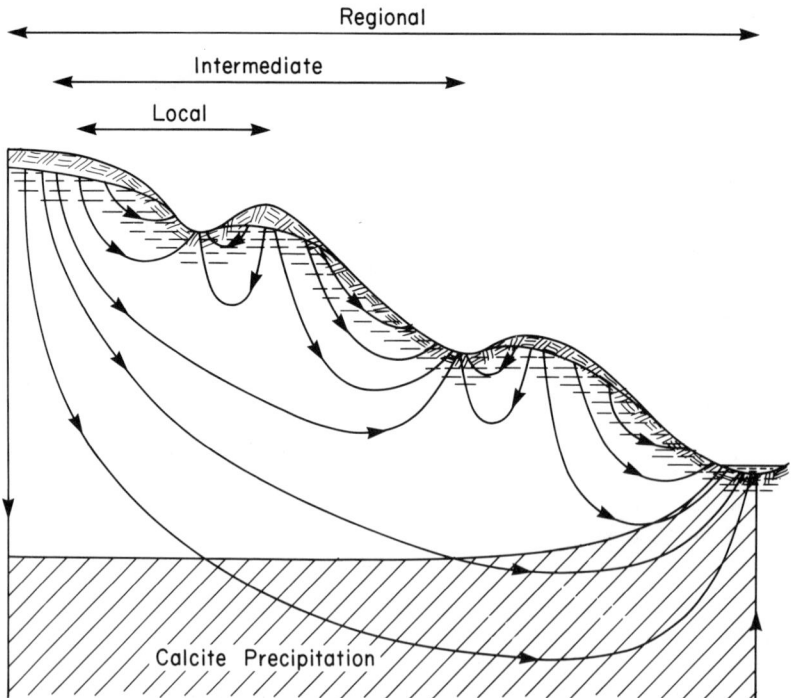

Fig. 3.8 Schematic picture of groundwater flow lines in a basin where calcite precipitation takes place due to weathering of plagioclases.

minerals, for example, rocks like granites, gneisses and schists. It is unlikely to take place in shales and sandstones unless they contain feldspars.

The effect of mineral transformations on oxidation–reduction conditions
There are a number of primary minerals which contain ferrous iron; hornblende and biotite are two examples. They are mostly considered to be fairly easily weathered. Since ferrous iron when released will rapidly convert into ferric hydroxides (in the presence of oxygen) it is thought that this oxidation helps to break up the protective coating on the surfaces of these minerals. In this way they would aid a more rapid removal of cations and hydroxyl ions than is possible in the case of feldspars.

The oxidation of ferrous iron into ferric iron must use dissolved oxygen since the system is closed with respect to gases. The oxidation can be illustrated by the reaction

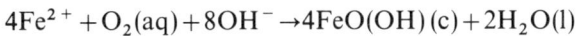

$$4Fe^{2+} + O_2(aq) + 8OH^- \rightarrow 4FeO(OH)(c) + 2H_2O(l)$$

This is an interesting reaction since it shows that hydroxyl ions are consumed at the same rate as ferrous iron is oxidized. Ferrous iron released in mineral transformation must be accompanied then by an equivalent amount of hydroxyl ions which of course disappear when the ferrous iron is converted into ferric hydroxides. There will thus be no change in pH when ferrous iron is released and oxidized.

As time goes on, the oxygen storage will be depleted, initiating reducing conditions. The consequences of this situation are best illustrated by stability diagrams. In Fig. 3.9 the negative logarithm of the partial pressure of oxygen, pPO_2 is shown as a function of pH for the various borders between phases and/or components. One can easily transfer this type of stability diagram into E_h–pH diagrams used frequently to illustrate stable phases and components in oxidation–reduction systems. Consider first the stability diagram in Fig. 3.9 of ferrous–ferric iron and their hydroxides. Total carbon dioxide is set at 0.1 mmol l^{-1} and total iron at 1 mmol l^{-1}. It is seen that goethite dominates the region while siderite, $FeCO_3$, occupies a fairly small sector of the diagram, i.e. between pH 6.7 and 8.6. Ferrous ions are found below pH 6.7 at very strongly reducing conditions. The dashed line illustrates the time development of the system as fayalite, Fe_2SiO_4, is weathered, releasing ferrous ions and hydroxyl ions; the latter are consumed during oxidation to ferric iron, except under strongly reducing conditions when part of the iron will be present as ferrous ions. The pH increase when ferrous ions are allowed to be present is quite strong hence keeping ferrous ion concentrations low. If there are buffering substances present however, the pH change will be small.

When sulphate is present is may supply oxygen by forming ferrous

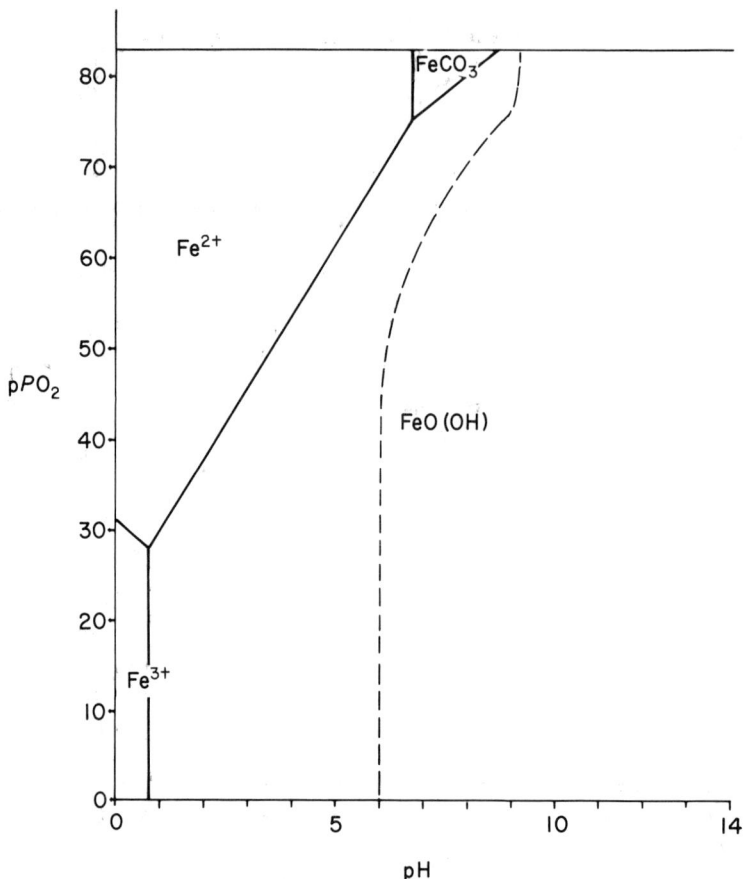

Fig. 3.9 The stability diagram of the iron–water–oxygen gas system, similar to that in Fig. 2.7 but with a line illustrating the time development of pH in the system when fayalite, HFe_2SiO_4 is weathered.

sulphide, FeS_2, also known as pyrite. In the presence of sulphur, the ferrous–ferric system in Fig. 3.9 becomes rather complicated, as shown in Fig. 3.10. In this system, there are 1 mmol l^{-1} of iron and 3 mmol l^{-1} of sulphur. There is thus an excess of sulphur above that required for formation of pyrite. (The carbon dioxide partial pressure is kept below 0.001 atm.)

The diagram shows that pyrite will form at fairly low partial pressures of oxygen. There is a wedge of elementary sulphur also at quite low pH values. When all the sulphur is reduced, at a pH of greater than 4, the remainder will consist of hydrogen sulphide and sulphide ions.

74

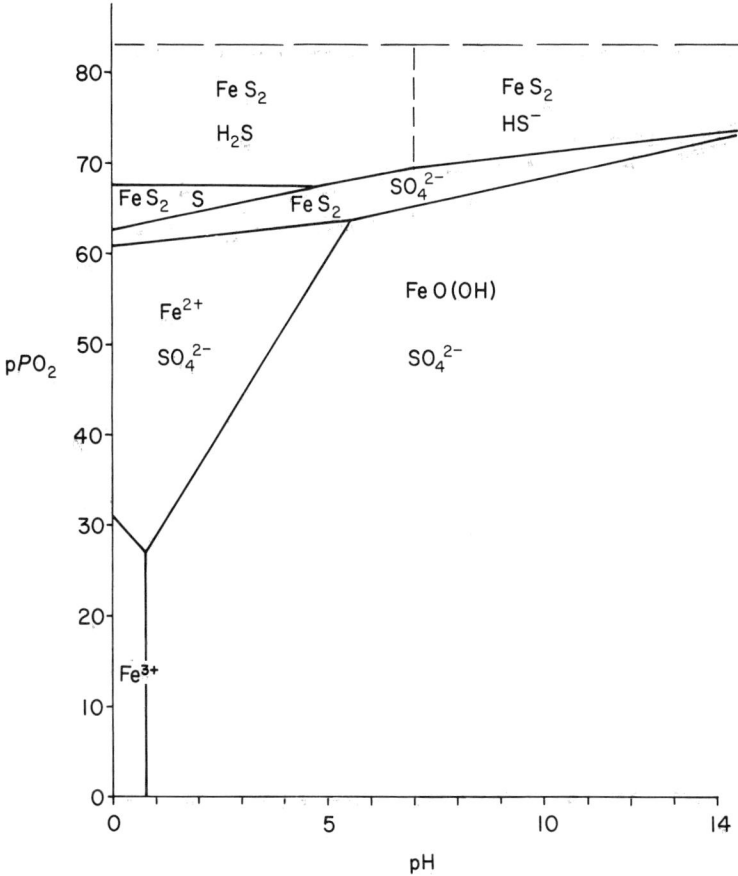

Fig. 3.10 The stability diagram for ferrous–ferric iron in the presence of sulphur. Total iron is 1 mmol l^{-1} and total sulphur 3 mmol l^{-1}. The carbon dioxide partial pressure is less than 0.001 atm.

For groundwater to reach the top of the diagram, there must be organic substances present of strongly reducing character. Only then will the hydrogen sulphide and sulphide ions be stable. If ferrous iron is the only reducing component present it cannot turn sulphate into sulphide ions simply because it will be locked up in pyrite. However, provided the pH is high enough, ferrous ions can be oxidized to form goethite. The situation can be illustrated by considering the following reactions

$$Fe^{2+} + 2SO_4^{2-} + 2H^+ \rightarrow FeS_2(c) + H_2O(l) + 3.5O_2(g)$$

and

$$2FeS_2(c) + 5H_2O(l) + 7.5O_2(g) \rightarrow 2FeO(OH)(c) + 4SO_4^{2-} + 8H^+$$

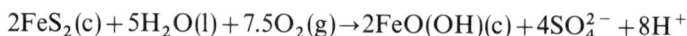

The first equation leads to the equilibrium

$$pPO_2 = 58.31 + 0.286pFe + 0.571pSO_4^{2-} + 0.571pH$$

being a line in the stability diagram between ferrous ion and pyrite, shown in Fig. 3.11 as the upper limit of the stability field for low pH values. The second reaction leads to the equilibrium

$$pPO_2 = 56.71 + 0.533pSO_4 + 1.067pH$$

the line separating FeS_2 and $FeO(OH)$ in the diagram. This line is the upper boundary of the system under consideration. In this case, it is only the ferrous iron that can consume oxygen, being slowly released through the weathering of primary minerals that contain it.

The stoichiometry of the reactions with sulphate is quite interesting. In the first reaction one ferrous ion consumes two sulphate ions, in this way releasing three-and-a-half oxygen molecules. This oxygen is then used to oxidize fourteen ferrous ions provided the pH is sufficiently high to keep goethite stable. Thus two sulphate molecules will take care of fifteen ferrous iron molecules, one of which will be tied up as pyrite. The ratio between pyrite and goethite will be 1:14 for any pH above 5. For lower pH values there will simply be only ferrous ions since there is no substance present to use up the oxygen released. If organic matter was present this could consume the oxygen and thus help in the formation of pyrite even at low pH values.

The oxidation potential E_h can be calculated for the FeS–FeO(OH) line. Since

$$E_h = 1.229 - 0.059pH - 0.01479pPO_2$$

and

$$pPO_2 = 56.71 + 0.533pSO_4^{2-} + 1.067pH$$

one obtains

$$E_h = 0.390 - 0.075pH - 0.008pSO_4$$

The following E_h values are worked out from this equation at selected pH values.

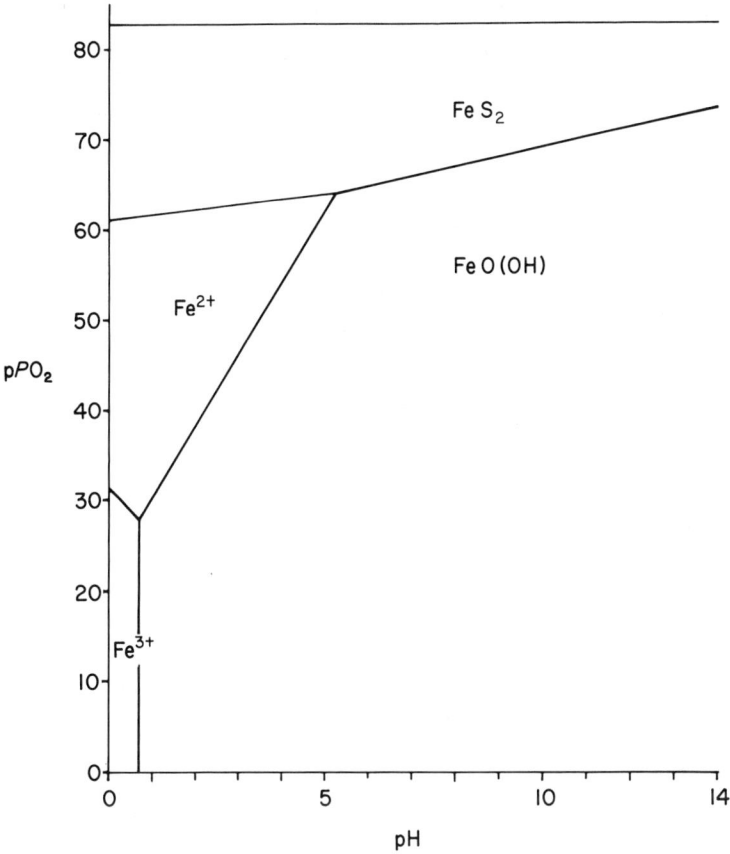

Fig. 3.11 The stability diagram for ferrous–ferric iron in the presence of sulphate, when the only reducing component is ferrous iron released in weathering. Note that the upper limit of the stability field in this case is set by the pyrite line.

pH	6	7	8	9	10
E_h	−0.086	−0.161	−0.236	−0.311	−0.386

A system containing pyrite and goethite is, of course, well buffered against changes in E_h.

The time development of the system discussed is fairly simple. The oxidation of ferrous ions to ferric ions in goethite does not affect the pH very much. Hydrogen ions are used up during the formation of pyrite from ferrous ions and sulphate, in addition to that used up when ferrous ions are released from silicates. The amount of pyrite formed is very small when compared to goethite formation and more important for pH is the release of sodium from silicates. Also, affecting the pH is the alkalinity produced during the oxidation of organic matter during sulphate reduction.

Ionic exchange
In a geologically homogeneous aquifer one would not expect ionic exchange to have any effect at all on the chemistry of water. There is certainly a fair supply of ion exchange material, particularly clay minerals of the Si–Al–Si type like montmorillonite which can act as ion exchangers. However, in a steady state condition all these ion exchangers have had ample time to equilibrate with the water. Therefore, there will be no change in the chemical composition of water due to ionic exchange.

Conditions in geologically heterogeneous aquifers may be quite different. In sedimentary rocks, marine deposits are often interlayered, forming a mixture of sandstones, limestones, and shales. The shales can have a considerable storage of sodium being deposited in equilibrium with sea water. This sodium is exchangeable. During the passage of calcium-rich, fresh water, sodium would be replaced by calcium. A calcium bicarbonate water would then be transformed into a sodium bicarbonate water. Such a process would continue until the whole bed has been equilibrated with the percolating water. When this stage is reached the chemistry of the water will return to its original state. The time for equilibration may be very long, tens of thousands or hundreds of thousands of years, depending on the exchangeable storage, the calcium concentration and on the flow-rate of water through the bed. Ionic exchange is a transient process, the effect of which will fade as time goes on.

Effects of saline formation water
Marine sediments will invariably contain sea water. During the regression of the sea (caused by land upheaval, for example) the sediments become a part of the terrestrial aquifers. A major part of such an aquifer can still be situated below sea level but, provided the fresh water aquifer possesses enough head, the formation water will gradually be pushed out. In deep aquifers this may take time, since density stratification helps to resist replacement. Molecular diffusion will create a diffuse front which slowly eats away the formation water. The process of replacement may even take millions of years.

As long as the natural circulation of water is undisturbed the contribution of formation water to discharging groundwater will be small. A high contribution would mean that the formation water would be washed out in a relatively short time. Only when a fairly small fraction of formation water is present in the discharging groundwater can the formation water last for a prolonged time. As an example, consider a 500 m thick sandy marine formation, able to store 50 m of water. The yearly recharge to the regional flow system is, say, 100 mm year^{-1} (taken over the whole catchment). Thus, in $50000/10 = 5000$ years the formation water would be replaced by fresh water. With 1 mm year recharge the time needed for replacement is 50000 years.

Saline formation water can also be present in pockets which are not connected to the rock aquifer. In such a case nothing will happen until somebody happens to drill a well into the pocket.

Effect of substances injected from great depths
It appears that gaseous exhalations take place in certain areas, particularly of carbon dioxide. Paces (1972) describes some springs in the Bohemian Alps which have unusually high bicarbonate concentrations. From chemical analyses he worked out partial pressures of carbon dioxide of about 1 atm. Since the root zone can be excluded as a possible source for this, he concludes that the origin is plutonic, i.e. it originates from great depths. This discovery is important although less spectacular contributions may take place in many aquifers, particularly in igneous rocks where tensile fractures may be very deep. As a matter of fact, water bearing fractures are frequently found along dykes of basic rock material. The deepest fractures are thus filled with basalt while less deep ones are open and filled with water.

Other constituents are also likely to emanate from great depths, for example, the transformation of hydrogen sulphide into sulphate when oxidized. It may also be the case for organic compounds.

3.5 Processes in groundwater discharge areas

3.5.1 *Definition and description of discharge areas*

Groundwater discharge takes place when groundwater flow is directed towards the groundwater surface in a phreatic aquifer. From this it follows that the aquifer suffers a loss of water in the discharge area. This loss can be accounted for as overland flow when the groundwater surface is at the soil surface or, more frequently, lost by evapotranspiration. In the second case the groundwater surface is either in the root zone or below the root zone at a distance close enough to allow capillary contact with it.

Overland flow in the strict sense will be rare in groundwater discharge areas except during heavy rainfall. The hydraulic conductivity of the root zone is mostly considerably higher than that of the intermediate zone. Hence, in groundwater discharge areas where evapotranspiration is unable

Principles and applications of hydrochemistry

to match the flow towards the surface, the streamlines of groundwater flow will converge in the root zone into a flow more or less parallel to the soil surface. Because of evapotranspiration the streamlines closest to the surface will end up on this surface. The deeper streamlines will reach the 'overland flow' area.

At low discharge rates, the groundwater discharge areas will be small while at high discharge they will be fairly extensive. During the spring flood in temperate climates, the local groundwater flow dominates while in mid-winter it is the turn of the intermediate or even regional groundwater flow. The discharge areas are then of limited size, close to the permanent drainage channels. This is of interest from the water chemistry point of view. Low basin discharges can be expected to have higher concentrations of soluble weathering products than high basin flows, which have a greater fraction of local groundwater flow. During rainfall, percolating rain partly mixes with the discharging groundwater and the area of overland flow increases.

Figure 3.12 shows schematically the features discussed. Obviously, vegetation will be favoured in the parts of discharge areas where the plant roots are able to extract water and nutrients from the discharging groundwater. The root zone may occasionally be submerged in water, for example during snow-melt and high rainfalls but this usually lasts only a few days and often revives the activity of the root zone. The turnover of organic matter in an aerated groundwater discharge area is quite substantial as compared with groundwater recharge areas. Adjacent to permanent drainage channels, the groundwater surface may be close to the soil surface most of the time, creating reducing conditions and, perhaps, even peat formation. The organic production, however, will still be comparatively high.

Thus, one can expect a considerably higher storage of soil organic matter in groundwater discharge areas than in recharge areas where conditions tend to exhaust the soil of cations and nutrients.

3.5.2 *Effects of evapotranspiration on dissolved salts*

Since part of the emerging groundwater in discharge areas will be lost by evapotranspiration there will be a corresponding increase in salt concentrations. The diurnal variation in evapotranspiration will thus cause a corresponding diurnal variation in the electrical conductivity of the streams into which groundwater flows, possibly with some time-lag depending on the width of the discharge area. Such periodic fluctuations have been observed in small streams (Calles, 1982) and even the diurnal variations in stream flow were detectable.

In semi-arid areas, groundwater discharge in alluvial plains often takes

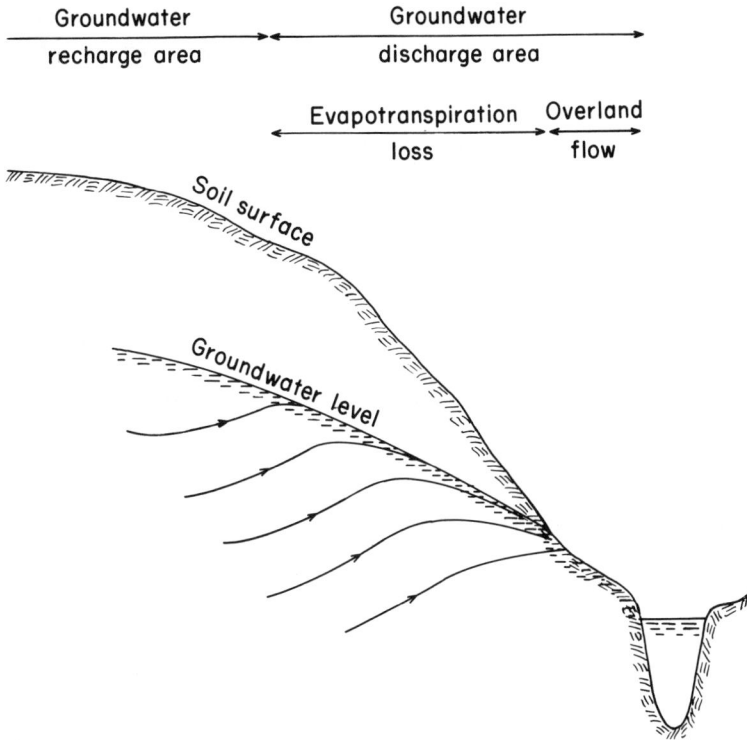

Fig. 3.12 Schematical picture of groundwater flowline patterns in a groundwater discharge area. Flow lines ending on the groundwater surface account for loss of water by evapotranspiration.

place in small depressions which consequently become salt marshes. Salts are thus transported into these depressions thus forming local groundwater flow systems.

3.5.3 *Calcite precipitation in groundwater discharge areas*

When groundwater is saturated with calcite the water will have a corresponding partial pressure of carbon dioxide. It is at this stage a 'closed system' with respect to carbon dioxide. When this water reaches a groundwater discharge zone it is brought in contact with the atmosphere and from then on it is an 'open system' with respect to carbon dioxide. Consequently, it has to equilibrate with the new conditions. The groundwater may then become undersaturated or supersaturated with calcite, depending on the new partial pressure of carbon dioxide in the discharge area. Furthermore, evapotranspiration may increase the calcium and bicarbonate concentrations, thereby also shifting the equilibrium conditions.

As to the effect of changing partial pressure of carbon dioxide this is simple as long as carbonate ions can be ignored. Recalling the equation

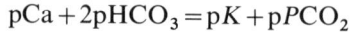

$$pCa + 2pHCO_3 = pK + pPCO_2$$

small changes in the variables can be approximated by the differentials. Using δ as an operator (denoting a small change) the following equation is obtained

$$\delta[Ca^{2+}]/[Ca^{2+}] + 2\,\delta[HCO_3^-]/[HCO_3^-] = \delta PCO_2/PCO_2$$

since pK is assumed to stay constant. Charge and mass balance require that $[HCO_3^-] = 2[Ca^{2+}]$ so that

$$(1/[Ca^{2+}] + 4/[HCO_3^-])\,\delta Ca^{2+} = \delta PCO_2/PCO_2$$

In the case $[HCO_3^-] = 2[Ca^{2+}]$ then

$$\delta[Ca^{2+}]/[Ca^{2+}] = (1/3)\,\delta PCO_2/PCO_2$$

Since $\delta PCO_2/PCO_2$ is a fraction, it is seen that for small changes, a 3% change in the partial pressure of carbon dioxide changes the calcium concentration by 1%. The last equation can be used by making small increments in the carbon dioxide partial pressure in each step of the calculation.

For more general and precise computations the equilibrium equation should be used with the condition

$$[HCO_3^-] = 2[Ca^{2+}] + E$$

where E is excess or deficit of bicarbonate with respect to the calcium present. When E is positive there are more bicarbonate ions present than are needed to balance the calcium charges. This may be due to albite weathering which produces sodium bicarbonate. Similarly, when E is negative there is insufficient bicarbonate present to balance the calcium charges. This may be due to solution of gypsum or oxidation of pyrite along the path of the groundwater, producing sulphate ions. Also in many regions, the deposition of sulphate from the atmosphere is a likely source.

Figure 3.13 shows the relationship between the concentrations of calcium and bicarbonate in a discharging groundwater which adjusts to a new partial pressure of carbon dioxide on contact with soil air. If a groundwater with a partial pressure of carbon dioxide of 0.01 atm emerges at the soil surface with a partial pressure of 0.0003 atm, then more than 2/3 of the calcium will precipitate. The amount precipitated is fairly independent of E,

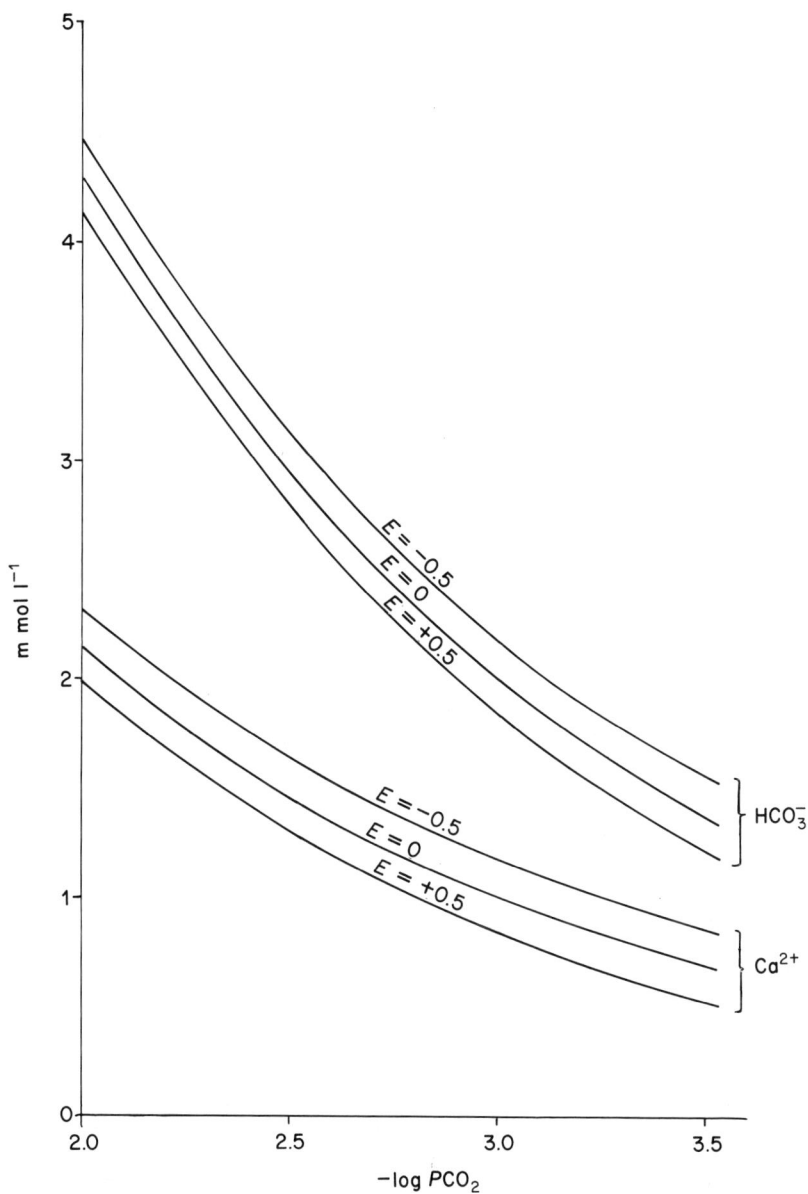

Fig. 3.13 The effect of changes in the partial pressure of carbon dioxide on discharging groundwater in equilibrium with calcite. E is excess bicarbonate over that required to balance the calcium ions. All concentrations in mmol l^{-1} and pressures in atm.

the excess bicarbonate over that required to match the electrical charges of calcium.

The effect of evapotranspiration on concentrations of calcium and bicarbonate can also be studied with the equation expressing equilibrium conditions and charge and mass balance, keeping the partial pressure of carbon dioxide constant but varying E (the excess bicarbonate). Since E is due either to cations, like sodium, or anions, like sulphate, its mass has to be conserved during evapotranspiration. This means that if 50% of the water is lost by evapotranspiration then E has to double. Working out the concentrations of calcium and bicarbonate for a number of E values makes it possible to reconstruct the relation between the concentrations and the fraction of water lost by evapotranspiration. If E_0 is the initial excess of bicarbonate then $1 - E_0/E$ is the fraction of water lost by evapotranspiration. When $E_0 = 0$, then there will obviously be no effect of evapotranspiration on the concentrations of calcium and bicarbonate, save for the small changes due to alterations in the ionic strength of the water. Calcium and bicarbonate will be removed in equivalent proportions during the evapotranspiration. The change in concentrations of calcium and bicarbonate with the fraction of water lost by evapotranspiration is shown in Fig. 3.14. The values were calculated using a solution initially at equilibrium with a partial pressure of carbon dioxide of 0.01 atm at a temperature of 5°C (i.e. assuming it was an ideal solution). The effect of concentrations on activity coefficients can be easily put in.

Evapotranspiration in groundwater discharge areas seems to be an efficient process for changing the water chemistry. A calcium bicarbonate water can be converted into either a sodium bicarbonate water or a calcium sulphate water, depending on the nature of the excess bicarbonate.

3.5.4 *Oxidation–reduction processes in groundwater discharge areas*

The groundwater flowing into a discharge area may or may not contain dissolved oxygen. The local groundwater flow usually carries well aerated water and nothing particular will happen when it emerges unless there are reducing substances in the soil where it enters. For groundwater levels immediately below the root zone there is no particular oxygen demand, but if the groundwater table is within the root zone, oxygen consumption will occur. It is probably of importance from the point of view of plants that the root zone can be partly or wholly submerged in a groundwater discharge area, yet there is sufficient dissolved oxygen in the emerging groundwater to maintain normal biological activity. If the discharging groundwater is low in oxygen, the submerged part of the root zone may become strongly reducing with peat formation as a consequence.

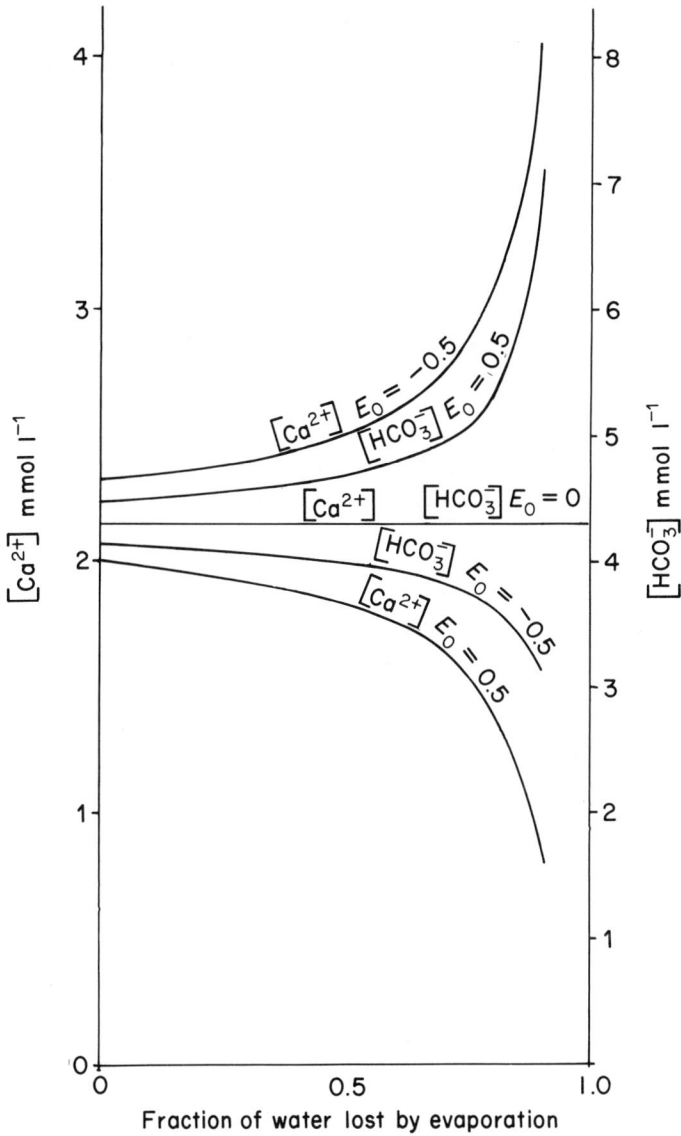

Fig. 3.14 Changes in the concentrations of calcium and bicarbonate in discharging groundwater due to loss of water by evapotranspiration E_0 is initial excess concentration of bicarbonate over that required for balancing the calcium concentration. The partial pressure of carbon dioxide is 0.01 atm and temperature 5°C.

The regional groundwater flow is likely to carry water which is oxygen free but the outflow areas of it are minimal and may even be part of drainage channels. Hence, no special ecological problems can be foreseen.

Discharging groundwater may contain appreciable concentrations of ferrous ions and manganese. The reason for this could be the dissolution of iron from B horizons due to high groundwater tables at some time in the past. Obviously, such processes cannot go on forever because the available storage of iron would be exhausted after some time and climatic changes are slow. Hence, if such dissolution of iron has taken place in the past, the groundwater must have been very slowly replaced and should therefore be part of the regional, or perhaps intermediate, flow systems. Springs with precipitation of ferric oxyhydroxides are, however not very common. The present groundwaters in the temperate regions of the Northern hemisphere containing ferrous iron were most likely formed at the end of the so-called 'climatic optimum' when peatland areas started to develop due to rising groundwater levels, thus releasing iron from submerged B horizons of podsols.

Ferrous ions are oxidized into ferric iron on contact with air and will appear as more or less crystalline goethite or lepidocrucite (another oxyhydroxide). The stability conditions for these compounds are seen in Fig. 2.7. The process of oxidation of ferrous iron can be written

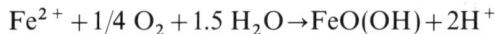

$$Fe^{2+} + 1/4\ O_2 + 1.5\ H_2O \rightarrow FeO(OH) + 2H^+$$

This reaction produces hydrogen ions and therefore the oxidation of ferrous iron brings about a lowering of the pH.

3.5.5 *Ionic exchange in groundwater discharge areas*

The soil in groundwater discharge areas will in general have a high cation exchange capacity. However, as long as the groundwater discharge can be considered to be a stationary process the ionic exchange reservoir will be in equilibrium with the groundwater flowing through it, taking into account also the increased concentration caused by evapotranspiration. There will be fluctuations in the concentration levels because of variations in evapotranspiration rates and in groundwater discharge rates, not to mention the effect of occasional rainfall. However, because the exchangeable ions act as a reservoir, ratios of ions tend to be preserved even if concentrations fluctuate. Apart from this, the composition of exchangeable ions is largely determined by the composition of the groundwater which has been subjected to the effects of evapotranspirations (as discussed previously).

3.5.6   *Dissolution of organic matter in groundwater discharge areas*

Since discharging groundwater must pass through the root zone it is inevitable that some organic matter is carried in solution and into the drainage channels. The degree of dissolution, however, depends much on the soil and the composition of the groundwater. If it is of the calcium bicarbonate type, the organic matter in the root zone acting as ion exchangers will probably be fairly saturated by calcium. In this case, organic matter is less prone to enter into solution than when it is saturated with sodium ions, i.e. a sodium bicarbonate water is likely to dissolve more organic matter than calcium bicarbonate water.

Dissolution is favoured by low electrolytic concentrations. When a fairly concentrated sodium bicarbonate water enters and saturates the cation exchange complexes, dilution by rainfall can cause a considerable dispersion of soil organic matter and strongly colour the water in streams. The dissolution of organic matter in soils is, to a large.extent, a dispersion of colloidal humic acids. It is not really a true solution except for short-lived organic acids. The dispersion depends strongly on the valency and ionic size of the exchangeable cations of the humic acids. Strongly hydrated monovalent ions like sodium favour dispersion of these acids.

3.6   Processes in lakes and water courses

3.6.1   *Definitions*

Lakes are considered to be any kind of surface water storage in that they are parts of drainage systems. This does not require any kind of outlet. A lake that receives water and loses it by evaporation is still a part of a drainage system. A water course can be anything from a small rivulet to a large river as long as it is part of a drainage system.

3.6.2   *Chemical processes in surface waters*

Calcite precipitation
The partial pressure of carbon dioxide in the groundwater approaches that of the atmospheric carbon dioxide, therefore, when the discharging groundwater is calcite saturated there will be a strong tendency for precipitation of calcium carbonate in lakes and streams. Also, a rise in temperature may aid precipitation because the solubility of carbon dioxide decreases with increasing temperature.

Much of the precipitation of calcium carbonate will be aided by the lower fauna in the water, for example, molluscs and foraminifera species. The calcareous shells they build up will, in general, have the solubilities of calcite or aragonite so that ordinary criteria on calcium carbonate saturation can be applied. Precipitated material will be incorporated into the bottom sediments of surface waters.

Precipitation of silica
Certain algal species form shells and incrustations of silica, SiO_2. This can effectively remove amorphous silica from the water, incorporating it with the sediments. Some reflux from the sediments will take place but in general there is a depletion of silica as compared to groundwater. Silica shells are thus an important part of bottom sediments.

Precipitation of aluminium and ferric oxides; suspended matter
Aluminium and iron in surface waters are predominantly present as humic acid complexes. In lakes they will be found in the suspended matter fraction. As such they follow the behaviour of colloidal matter, i.e. ageing, coagulation and sedimentation. It is instructive to note that a lake with a large turnover time for water (equal to volume/flow rate) has low concentrations of suspended matter and consequently very clear water. The colloidal matter brought in has ample time to age, increasing the sizes of suspended particles and allowing them plenty of time to settle to the bottom. Admittedly, this also removes nutrients like phosphorus.

3.6.3 *Effects of biological processes*

The biological processes in surface waters are fairly complicated when considered in detail. To begin with, carbon dioxide is converted into organic matter; this is then decomposed in steps following an intricate pattern in which a large number of species participate. The assimilation stage can be simplified by the reaction

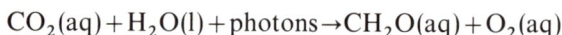

$$CO_2(aq) + H_2O(l) + photons \rightarrow CH_2O(aq) + O_2(aq)$$

The process consumes carbon dioxide and produces oxygen. In tropical climates the process is more or less constant throughout the year balanced by an equally constant rate of decomposition which uses up oxygen and produces carbon dioxide. The actual rate of assimilation is, however, set by the availability of dissolved phosphorus and nitrogen compounds. In shallow lakes, bottom vegetation seems to be more important than phytoplankton whereas in deep lakes the reverse is true. The phytoplankton will be eaten by zooplankton which then serve as food for higher animals. During this cycle, suspended matter is formed and sedimentation of some organic material (including some nutrients) is inevitable. In deep lakes there is thus a gravitational transport of organic matter in to depths where assimilation is no longer possible because of lack of light. A net loss of organic matter and nutrients thus takes place leaving a corresponding excess of oxygen formed during the assimilation and depriving the atmosphere of the equivalent amount of carbon dioxide.

The gravitational transport of organic matter creates vertical variation in carbon dioxide, oxygen and nutrients. Most of the suspended, slowly sinking organic matter will be decomposed into carbon dioxide with the release of nutrients as a consequence. Hence, the carbon dioxide and nutrient concentrations will increase with depth while oxygen decreases. The molar sums of the concentrations of oxygen and carbon dioxide will, however, be nearly conserved. Actually if the respiratory quotient (carbon dioxide produced)/(oxygen consumed) is r then the sum $O_2 + CO_2/r$ will be a conservative property like, for example, the chloride concentration. Vertical mixing will, of course tend to decrease vertical gradients which means that carbon dioxide and nutrients are brought back to the surface while oxygen is brought down. In large lakes more or less permanent current systems can modify the vertical distributions so that maxima and minima can appear at certain depths.

In temperate climates there will be a seasonal variation of both temperature and light. Consequently, the assimilation rate is varying throughout the year. The situation is complicated by the variation in temperature stratification within the lake. During the summer, the well-mixed upper part of the lake has a comparatively high temperature; this rests on a mass of much cooler water. Consequently there is a stable density stratification which practically isolates the deeper parts. However, these deeper parts receive organic matter through sedimentation which uses up the oxygen from the deeper water. In the autumn, surface water cools until a point is reached when water is at 4°C and a complete mixing of water takes place. During winter an ice cover will effectively put a lid on the water preventing any gas exchange. In addition, the cooling again causes a stable density stratification. Decomposition of organic matter continues although at a slow rate releasing carbon dioxide and nutrients and reducing the oxygen supply. Reducing conditions may occur if there is an ample supply of decomposable organic matter in bottom sediments.

When spring arrives the ice sheet disappears opening up the gas exchange between the lake and the atmosphere. At 4°C, the water mass becomes completely mixed and the scene is set for a new assimilation phase.

3.6.4 *Concluding remarks*

Except for the possible precipitation of calcium carbonate, lakes and streams do not influence the water chemistry to any remarkable degree. A complete mixing of all discharging water takes place creating a homogeneous chemical composition of the water. Silica and nutrients like phosphorus and nitrogen compounds, are depleted due to sedimentation of organic material. The water has a fairly high concentration of organic matter and is itself in approximate gaseous equilibrium with the atmosphere.

References

Albertsen, M. (1977) Labor- und Felduntersuchungen zum Gasaustauch zwischen Grundwasser und Atmospäre über Natürlichen und Verunreinigten Grundwässern. Dissertation Kiel 1977, 145 pp.

BAPMON (1981) *BAPMON data for 1978*, Joint WMO-EPA publication, National Climatic Center, Federal Building, Ashville, N.C. 28801.

Busenberg, E. and Clemency, C. V. (1976) The dissolution kinetics of feldspars at 25°C and 1 atm CO_2 partial pressure, *Geochim. et Cosmochim. Acta*, **40**, 41-9.

Calles, U. M. (1982) Diurnal variation of electrical conductivity of water in a small stream, *Nordic Hydrology*, **13**, 157-64.

Eriksson, E. (1952a) Composition of atmospheric precipitation. I. Nitrogen compounds, *Tellus*, **4**, 215-32.

Eriksson, E. (1952b) Composition of atmospheric precipitation. II. Sulfur, chloride, iodine compounds. Bibliography, *Tellus*, **4**, 280-303.

Eriksson, E. (1959) The yearly circulation of chloride and sulfur in nature; meteorological, geochemical and pedological implications. Part I, *Tellus*, **11**, 375-403.

Eriksson, E. (1960) The yearly circulation of chloride and sulfur in nature; meteorological, geochemical and pedological implications. Part II, *Tellus*, **12**, 63-109.

Eriksson, E. (1966) Air and precipitation as sources of nutrients, in *Handbuch der Pflanzenernährung und Düngung*, Springer-Verlag, New York, Vol. 2, pp. 774-92.

Foster, S. S. D., Bath, A. H., Farr, J. L. and Lewis, W. J. (1982) The likelihood of active groundwater recharge in the Botswana Kalahari, *J. of Hydrology*, **55**, 113-36.

Hingston, F. J. (1958) The major ions in Western Australian rainwaters, *CSIRO Div. Report Bureau of Soils*, **1**, 1-11.

Hutton, J. T. and Leslie, T. J. (1958) Accession of non-nitrogeneous ions dissolved in rainwater to soils in Victoria, *Australian J. of Agric. Research*, **9**, 492-507.

Jacks, G. and Sharma, V. P. (1982) Hydrology and salt budget in two tributaries of the Cauvery River, India, Association Franco-Suédois pour la Recherche, (AFSR) report No. 41, 155-165, Stockholm.

Luce, R. W., Bartlett, R. W. and Parks, G. A. (1972) Dissolution kinetics of magnesium silicates, *Geochim. et Cosmochim. Acta*, **36**, 35-50.

Paces, T. (1972) Flux of CO_2 from the lithosphere in the Bohemian Massif, *Nature Physical Science*, **240**, 141-42.

Tamm, O. (1940) *Northern Swedish Forest Soils*, Norrlands Skogsvårdsförbud Förlag Stockholm, pp. 121–22.

Further reading

Jacks, G. and Sharma, V. P. (1982) Hydrology and salt budget in two tributaries to Cauvery River, India, AFSR Report No. 41 (Stockholm), 155–65.
Troedsson, T. and Nykvist, N. (1973) Marklära och markvård (Soil science and soil management), Almqvist and Wiksell, Stockholm.

4

Models of reservoirs and the flux of chemical constituents in basins

In this chapter some models for the storage and fluxes of chemical constituents are discussed with a view to assessing their possible applications. In the past hydrochemical modelling focused on mixing processes in lakes and streams. In groundwater flow the emphasis was on longitudinal mixing, i.e. mixing in the direction of flow.

4.1 Concepts and definitions

The turnover of chemical substances in basins can be pictured by models using boxes or compartments. A model is a simplified picture of nature, constructed for a particular purpose. Thus, there is never a 'complete model' since this would be nature itself. A model disregards the complex details of nature which are considered to be irrelevant for the problem at hand. Statistical averages over areas and time segments are used as variables and parameters. The averaging may call for so-called parameterization of processes which cannot be included in the model for technical reasons. This is the case with mixing which is basically due to a complicated and varying velocity distribution coupled with molecular diffusion. The space and time resolution needed to picture such a process is prohibitive. The process is therefore described in terms of a uniform velocity and a diffusion coefficient which is very much larger than the molecular one. The numerical value of such a parameter is in most cases impossible to assess independently from the model. It is usually adjusted so that, within

certain limits, the model agrees with the observations. This parameterization can work well for prediction of events provided the input variables
can be predicted. However, it will fail to predict the effect of changes in the
system, such as the human influence in basins of clear-cutting and land
drainage.

The averaging referred to can be extended over entire basins. In a
conceptual model the basin is divided into parts which are known to
function differently. In preceding chapters the terms 'root zone', 'intermediate zone' and 'saturated zone' were used. In a conceptual model, these
can be represented as separate units. In a lumped model each unit is the
result of averaging over the entire basin. The root zone thus represents the
entire volume of the root zone in the basin. In a distributed model, each unit
represents selected sub-areas of the basin and if, in addition, the model is
conceptual the selected sub-areas must be sub-divided into a root zone part,
an intermediate zone part and so on. A distributed conceptual model
requires obviously much more information on parameters than a lumped
model. On the other hand topography can be used to help obtain
independent estimates of parameters which otherwise have to be assessed
by observations in calibrating the model. Hence, distributed models have a
greater predictive power than lumped models.

In hydrological models, the dynamic behaviour is of interest for
predicting events resulting from variations in input. The feature of the
models which makes this possible is called 'response'. This is made possible
by the fact that these models possess a 'response function'. However, a
considerable part of the response is response to pressure changes in the
saturated zone. Studies using stable isotopes of oxygen and hydrogen (^{18}O
and deuterium) show convincingly that the increase in river discharge in a
basin because of rainfall is due to groundwater being forced out. Thus, the
response function of river discharge models cannot be used to predict the
change in chemical composition of the river water due to rainfall. The basic
function required can be called the 'transit time distribution function' (or
alternatively the 'residence time distribution function') of groundwater in
the basin. The transit time distribution is the cumulative distribution of
travel times of water molecules through the basin. This function can be
assessed experimentally with the aid of a tracer substance with the same
chemical properties as water (for example, tritiated water). Another
possibility is to estimate this function when groundwater flow pattern and
hydraulic conductivities are known in some detail. With that much
knowledge at hand the use of a distributed model would automatically take
over the role of the transit time distribution function.

There are some possible simple 'models' for the transit time distribution.
One is the so-called exponential model suggested by Eriksson (1958)

93

considering actual groundwater flow patterns, the general shape of which is shown in Fig. 4.1. The exponential distribution is set by one parameter as the function $1 - e^{-\tau/\tau_0}$ where τ_0 can be called the 'mean transit time' or 'mean residence time'. More often, however, three parameters would be needed represented by $a(1 - e^{-\tau/\tau_1}) + (1 - a)(1 - e^{-\tau/\tau_2})$ where τ_1 would be fairly small – a matter of a few months or years – while τ_2 would be tens or hundreds of years or even larger. The actual values of the parameters could be estimated independently.

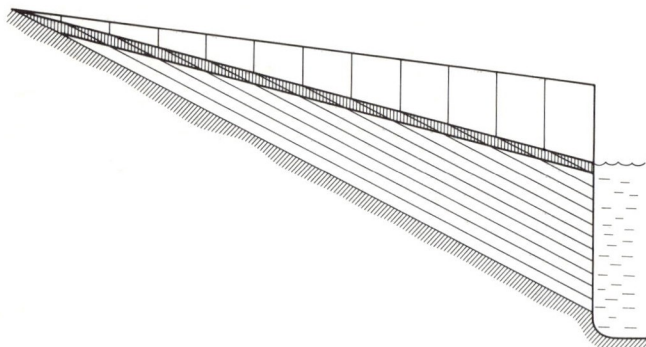

Fig. 4.1 Simplified picture of flow lines of water in a basin as a basis for the simple exponential model for describing the transit time distribution of flow pattern in an approximate way.

Chemical substances (other than chloride and nitrate) do not travel at the same speed as water due to ionic exchange or sorption on to mineral surfaces. Therefore, a function to describe the fraction of an ion in solution is also needed. This is usually called the absorbtion isotherm and is often based on empirical data. The apparent velocity of a substance where only the fraction α is in solution, is α times the velocity of water.

Chemical substances can be added or subtracted from groundwater as described in the previous chapter. Hence, any equation describing the flow rate of chemical substances must contain a source term giving the rate of addition or subtraction when the source strength is negative.

It is obvious that hydrochemical modelling is far more complicated than river discharge modelling. It requires more information from the basin than a river discharge model in order to be meaningful and it also requires a good understanding of hydrochemical processes.

Lumped conceptual models of mean fluxes and reservoir contents are useful in order to obtain an overview of the chemical system in a basin and the quantities involved. For the reservoirs, one can work out a turnover time, that is, the reservoir content divided by the mean flow rate of the substance flowing into the reservoir. When the transit time distribution for a component can be described by a simple exponential function then the mean transit time and the turnover time are identical.

4.2 A conceptual lumped budget model for groundwater recharge areas

As the heading implies this model pictures only the groundwater recharge parts of a basin. Figure 4.2 shows the arrangement of reservoirs and connecting flows. There is an atmospheric reservoir and a vegetation reservoir, the latter including living roots in the soil. These reservoirs are intimately tied to the hydrological units, i.e. the root zone, the intermediate zone and the saturated or groundwater reservoir. The flow pattern is quite general and can, with careful consideration apply to any chemical substance dissolved in water as well as to water itself.

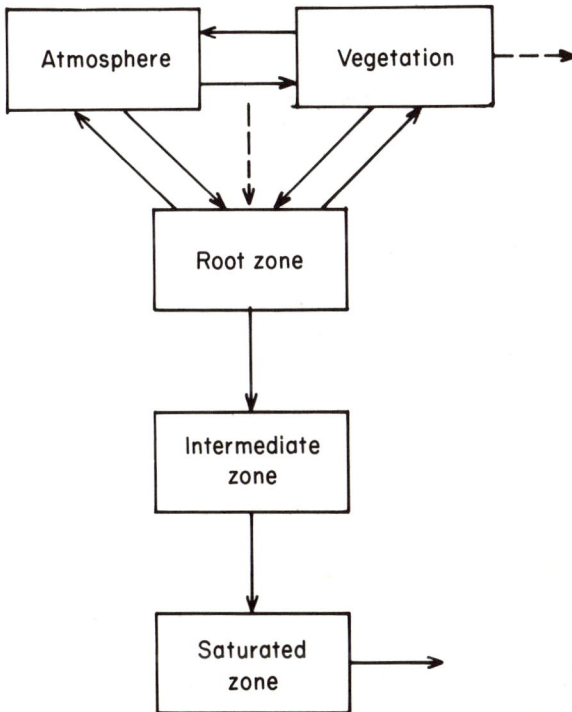

Fig. 4.2 A lumped model of reservoirs and fluxes related to the groundwater recharge area of a basin.

Principles and applications of hydrochemistry

In Fig. 4.3 figures are given of the reservoir contents and flow rates of chloride taken from coniferous forests in central Sweden. Figure 4.4 shows the reservoirs and flow rates of calcium for the same location. The turnover times are also shown in these diagrams. This representation of the major chemical features of a basin is certainly of great interest since it is the first step on the road to the construction of a model which can be run on a time

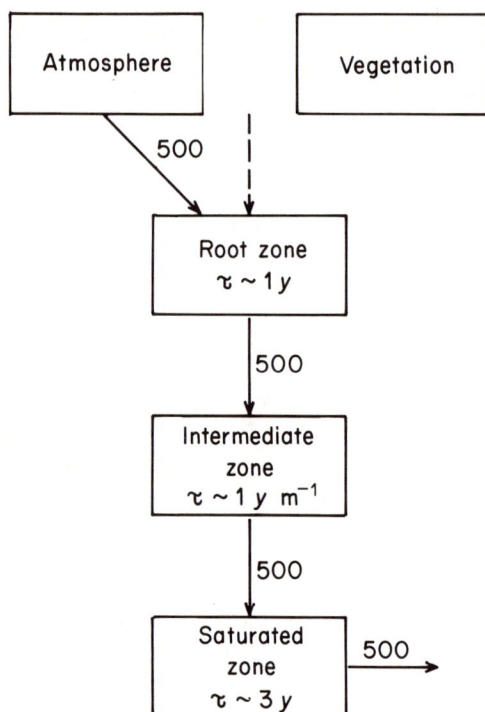

Fig. 4.3 This is a similar model to that in Fig. 4.2 but now applied to chloride in a coniferous forest in the middle of Sweden.

basis. The turnover times can then serve as parameters. One can most likely get some pertinent information for instance on the effect of acid rainfall, since hydrogen ions will replace calcium and other cations in the root zone, provided that the anions sulphate and nitrate are preserved. The effect of clear-cutting a forest is more problematic unless it is known how it affects the root zone storage of organic matter. Alternatively, a separate set-up of reservoirs and fluxes for carbon could be used in order to get a more complete picture of the effect of clearing. Calibration can be made by using

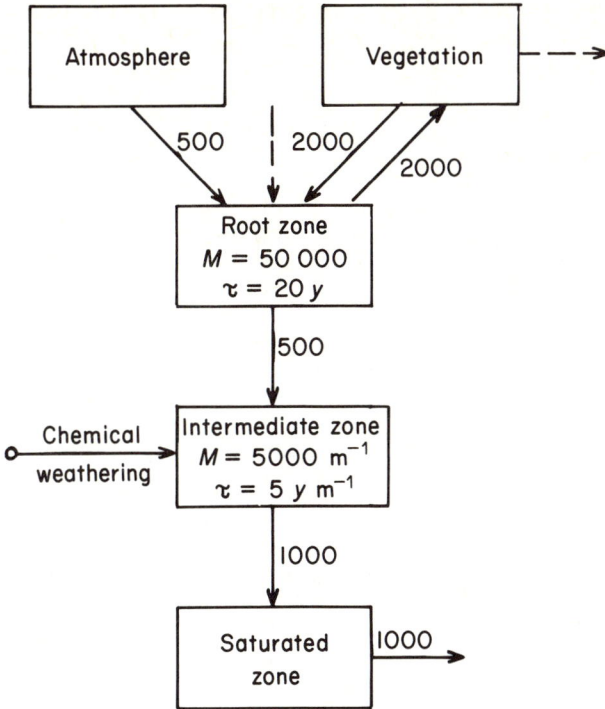

Fig. 4.4 This is a similar model to that in Fig. 4.2 but now applied to calcium in a coniferous forest in the middle of Sweden.

empirical data whenever possible. Fitted parameters are useless in this context.

4.3 A conceptual lumped budget model for groundwater discharge areas

Figures 4.5 and 4.6 show the result of applying this model to chloride and calcium. The groundwater now flows directly into the root zone and with it a considerable flow of ions. The root zone in this case will have a considerable storage capacity; the outflow from the root zone forming surface water. From an ecological point of view, there is an enormous difference between the root zones in recharge and discharge areas. It would be quite feasible to use this framework as a model for predicting the effect of changing land use. Little can be done to the groundwater part of the flow without drastically changing conditions in the recharge areas. Clear-cutting may increase mineralization and release calcium but new vegetation

97

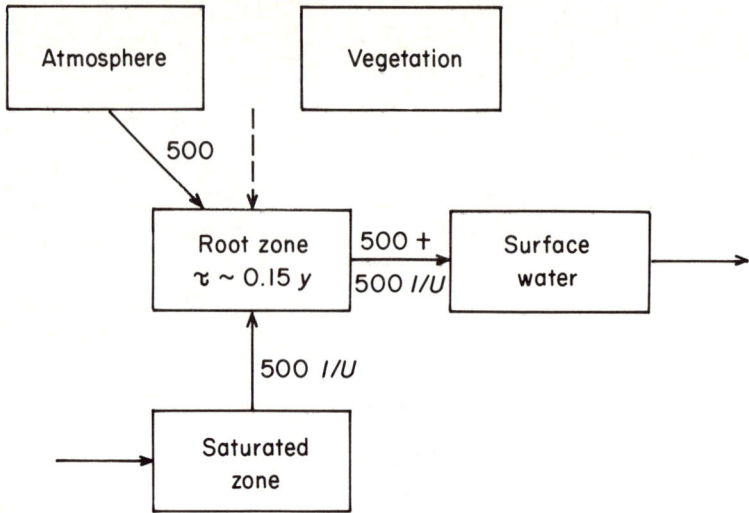

Fig. 4.5 A lumped model of reservoirs and fluxes related to groundwater discharge areas applied to chloride in a coniferous forest in the middle of Sweden. I/U is the ratio of recharge to discharge areas.

is soon established, hence it is doubtful if the effect would be noticeable.

The models in Figs 4.5 and 4.6 can most likely be used to simulate variations in salt concentrations with rainfall or snowmelt. For this purpose, they must be combined with the corresponding recharge area

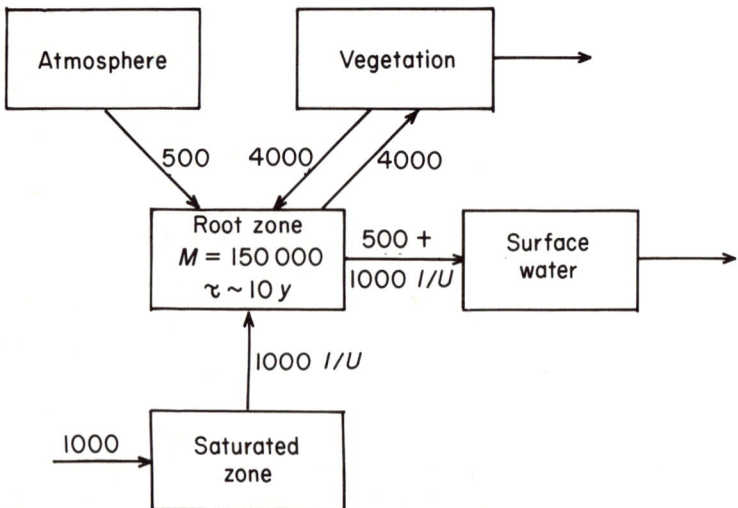

Fig. 4.6 A similar case to that in Fig. 4.5 but now applied to calcium.

models, since inflow of groundwater into the discharge model is the same as the groundwater outflow in the recharge model. The effect of rainfall is therefore to increase outflow of groundwater in the recharge model thus increasing also the flow to the root zone in the discharge model. Some mixing with rainwater takes place which affects the concentration of the surface water.

It is quite feasible to use the conceptual models discussed for simulation. These models also aid the interpretation of hydrochemical data from streams, and are of interest to biologists and pedologists in ecology and soil chemistry studies.

4.4 Distributed models and their limitations

A distributed model can utilize physical laws of water flow in the ground, that are impossible in the case of a lumped model. A distributed model consists of a number of sub-areas, each sub-area being treated as a reservoir or as a sequence of reservoirs which simulate, say, the root zone, although this is only true in a geologically homogeneous aquifer. There may be cases when vertical inhomogeneity calls for further divisions as when, for instance, a less permeable layer is present in the intermediate zone forming an aquitard, thereby permitting water saturation at two levels. In such cases there are both vertical and horizontal flows to consider in the upper saturated zone.

Figure 4.7 shows schematically divisions of a sub-area into conceptual volume elements in the simple case with root zone, intermediate zone and saturated zone. Only the saturated zone permits horizontal flow, the others transfer water only vertically. In a groundwater discharge area there would be no intermediate zone and the vertical flow would be directed upwards.

Fairly simple rules can be set up which govern the vertical flow. In the root zone the difference between rainfall and evapotranspiration will accumulate in the reservoir until field capacity is reached when it starts moving into the intermediate zone. The flow rates can be related to the water contents of the root zone and the intermediate zone. In a discharge area the root zone will receive water both from rainfall and upward flow from the saturated zone, the latter being set by the divergence of the horizontal flow. If the root zone becomes saturated, water will flow horizontally as in the saturated zone. As for the saturated zone, the driving force for the horizontal flow is proportional to differences in water levels between adjacent volume elements. In this way, the horizontal in- and outflows for each element are worked out from field data on the hydraulic conductivity and the depth of the saturated zone. When the horizontal flow components are added up, any difference is treated as flux divergence and

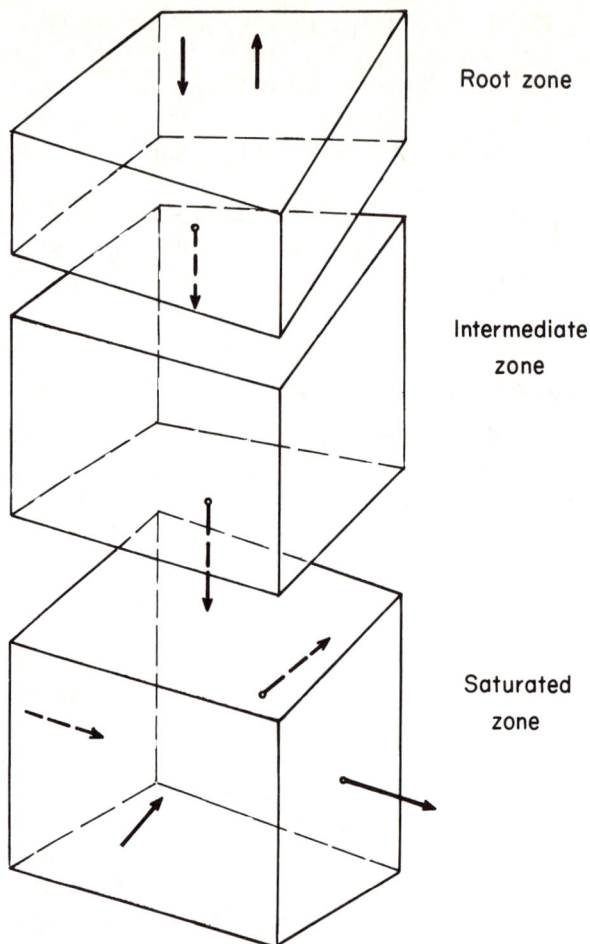

Fig. 4.7 Conceptual sub-division of a sub-area into volume elements. The horizontal flows in the saturated zone are proportional to the water level differences between the elements.

will increase or decrease the thickness of the saturated zone, or flow into the root zone as in a groundwater discharge area. The relation between thickness variation and flux divergence is given by the effective porosity, usually worked out from pump test data, together with hydraulic conductivity. The whole flow pattern in a basin can thus be simulated to give the topography of the groundwater surface. This can then be checked by observations and if the parameters determined independently in the field are reliable, then observed and simulated groundwater levels should show satisfactory agreement.

Mathematical groundwater flow models are available as computer programs. In one type, the 'finite difference models', the sub-areas are all of the same size forming a regular network. In the 'finite element models', the sub-areas can have any size and shape and can in this way be adjusted to the shape of the basin and its drainage network. This is not possible in a finite difference network.

As far as chemical constituents are concerned, a number of additional storages and flows have to be included such as uptake by vegetation, deposition by litter fall and release by organic matter decay. In a steady-state condition, ionic exchange need not be considered. However, calcite when present, must be taken into account, as well as the contribution of common cations from mineral transformations. Also, the possible precipitation of calcium carbonate in groundwater discharge areas must be watched. Field observations of the chemistry of the water aids reconstruction of its history and the parameters to be used are generally independently determined. Thus there are no formidable obstacles for steady-state solutions.

If, however, changes have to be incorporated into the model, additional information is needed about the fraction of each constituent in solution. However, if this is known there is no real problem. Some longitudinal mixing may have to be considered in the balance calculation of each saturated sub-area element. This dispersion can be treated as an exchange of water solution between adjacent reservoirs but certain parameters must be applied. There is no other way.

There is one obvious weakness of the model described above: it does not account for the various scales of flow (local, intermediate and regional) because the vertical properties have been integrated. The local scale will *always* dominate when considering the flow of water. However, in hard rock areas with primary minerals releasing most of the dissolved substances, transit times are more important since mineral transformations are slow but steady. Neglecting the regional flow within a basin can introduce considerable errors in the model behaviour with respect to chemical constituents. In a steady state, assuming that mineral transformations proceed at a constant rate, the total weathering rate depends on the total surface area of the minerals which are transformed and it must equal the rate at which released components (sodium, in particular) are transported out of the basin. Since the regional flow may occupy a much larger volume of rock than the local and intermediate, it will also supply more sodium to the streams than the local flow. In the case of calcium, the reverse may even be true if precipitation of calcite takes place. Potassium is always limited because of its involvement in the formation of some secondary layered minerals.

Thus, the distributed conceptual model is not always ideal for chemical constituents even if the water flow is represented correctly. A three-dimensional model would be the answer but there seems to be little evidence of this in groundwater hydraulics. A possible solution is to divide the saturated zone reservoir into two parts, a lower one with a constant volume and an upper one where divergence of flow is taken care of by adjusting the groundwater level. If hydraulic conductivity decreases with depth, then even several constant volume sub-area elements could be incorporated into the model. This refinement is not likely to improve groundwater flow patterns in general but should definitely improve the flow pattern of dissolved ionic components.

References

Eriksson, E. (1958) The possible use of tritium for estimating groundwater storage, *Tellus*, **10**, 472–78.

Further reading

Eriksson, E. (1971) Compartment models and reservoir theory, *Annual Review of Ecology and Systematics*, **2**, 67–84.

5

Environmental isotopes in hydrology and hydrochemistry

The term 'environmental isotopes' was proposed some time ago for a number of globally distributed isotopes of various elements. One of the environmental isotopes is tritium, or ^3H, of cosmic ray origin. During the hydrogen bomb tests in 1952 to 1962, it was released into the atmosphere in quantities which were enormous compared with the cosmic ray production. It spread globally and although it was consequently present everywhere in the environment, it can hardly be called 'natural'. The same is true for ^{14}C and a number of other radioactive isotopes which are not discussed here.

The stable isotopes of the water molecule, deuterium, ^{17}O and ^{18}O and the stable isotope of carbon, ^{13}C may also be considered in this category. The proportions of deuterium and oxygen isotopes in water vary due to fractionation processes during the evaporation and condensation of water in nature. They will be described in some detail in a later section. They have proved valuable as tracers, particularly in groundwater hydrology. The ^{13}C also displays fractionation effects useful sometimes for identification of the origin of carbon species in solution and in calcite. It is also used to correct ^{14}C data for fractionation effects during the past history of carbon.

The literature on environmental isotopes is quite extensive. Of special value is the *Handbook of Environmental Isotope Geochemistry* (eds Fritz and Fontes, 1980) and the monograph on the stable isotopes of water published by the International Atomic Energy Agency (eds Gat and Gonfantini, IAEA, 1981). This organization has also published a large number of proceedings from symposia, specifically on applications of isotopes in hydrology as well as other publications on the subject. Informative parts are also found in the *Guidebook on Nuclear Techniques in Hydrology* (IAEA, 1983). The presentation here will be limited to those environmental isotopes which have been found to be of particular value in groundwater investigations.

5.1 Radioactive isotopes of hydrogen, carbon and chloride

The radioactive isotopes of hydrogen, carbon and chloride are ^3H (tritium), ^{14}C and ^{36}Cl, respectively. They are formed by neutron capture mainly in the upper atmosphere by atmospheric oxygen, nitrogen and argon. The production rate of tritium averages about 0.25 atoms $(cm^{-2} s^{-1})$ and that of ^{14}C about 2.5 atom $(cm^{-2} s^{-1})$ while that of ^{36}Cl is very much lower. The radioactive half-lives are the following

tritium	12.43 years
^{14}C	5730 years
^{36}Cl	450000 years

Other radioactive isotopes like ^{32}Si and ^{35}S are cosmic-ray produced and may in special cases have hydrological interest.

5.1.1 *Tritium*

The concentration of tritium is usually expressed in Tritium Units, TU. 1 TU is one tritium atom to 10^{18} hydrogen atoms. The steady state concentration levels in atmospheric precipitation range from a few TU to about 20 TU, depending on the origin of the air moisture and its history. Fresh waters have about the same range. In the surface waters of the oceans the steady state level is generally low, less than 1 TU.

The tritium is determined by its β-particles emitted on radioactive decay. Because the radiation is extraordinarily weak, the monitoring of decay has to be made in counters filled with hydrogen gas or with the liquid scintillation technique usually on samples enriched previously by electrolysis.

Before hydrogen bomb testing took place, tritium was not much used in hydrological studies. The major pre-bomb work was published by Begemann and Libby (1957) and attracted great attention because of their hydrological interpretation of the data.

Considering the decay rate of tritium and a reservoir of water the inflow of tritium by rainfall has to be matched by the sum of radioactive decay in the reservoir and the outflow. Assuming rapid mixing of the groundwater reservoir, the outflow concentration always equals the average concentration in the reservoir. If C_i is the inflow concentration and C_σ the outflow concentration, the reservoir volume is V and the flow rate F, then in a steady state the tritium balance can be written

$$FC_i = FC_\sigma + \lambda C_\sigma V$$

where λ is the radioactive decay rate equal to 0.693/(half-life). For tritium, λ is thus 0.0558 year^{-1}. From the equation it is seen that

$$V = (F/\lambda)(C_i - C_\sigma)/C_\sigma$$

Hence, if C_σ is 50% of C_i and F is 200 mm year^{-1} then $V = 3540$ mm stored as, say, groundwater. The turnover time of water then becomes 17.6 years.

Strictly speaking, the equation above is only applicable when the transit time distribution in the reservoir is an exponential of the type $1 - e^{-\tau/\tau_0}$ (τ_0 being a parameter). As discussed in a previous chapter, this assumption can be used as a first approximation but the volume V is likely to be underestimated.

The hydrogen bomb tests in 1952 and onwards destroyed the possibilities of further use of naturally produced tritium in groundwater studies practically all over the globe. The new concentrations were many orders of magnitude higher than the natural ones. Although at present, concentration levels decrease with time it will be a very long time before concentrations are back to normal, if ever.

The bomb tests introduced a very large transient on the natural steady state. Hence, tritium decay as such cannot be used in the simple manner possible before the bomb tests. Yet, a transient state can be used in systems analysis to derive response characteristics. Similarly, it can be utilized in hydrology to derive transit time distributions. Hence, bomb tritium can be regarded as a tracer put into the continental water systems. By recording input by rainfall and output by river run-off and analysing the data with techniques used in systems analysis, much more information on the parameters on the reservoirs is obtained than when analysing a steady state situation. However, in order to utilize the transient effectively, data must be collected from the very beginning of the transient state. Apart from a few places, monitoring of tritium in rainfall and rivers started several years after the initial 1952 tritium injection into the atmosphere, and in many cases data on rivers exist only after 1962. There is only one river system with a fairly complete record – the Ottawa River in Canada.

Data were collected and analysed by Brown (1961) who applied the exponential model (one well mixed reservoir) with some success although there are noticeable deviations between the simulated data and the observations, indicating the inadequacy of the simple exponential transit time distribution approach. Eriksson (1963) tried later to analyse the Ottawa series by manual adjustment of the transit time density distribution and arrived at the density distribution shown in Fig. 5.1. It also shows the transit time density distribution for the simple exponential model used in

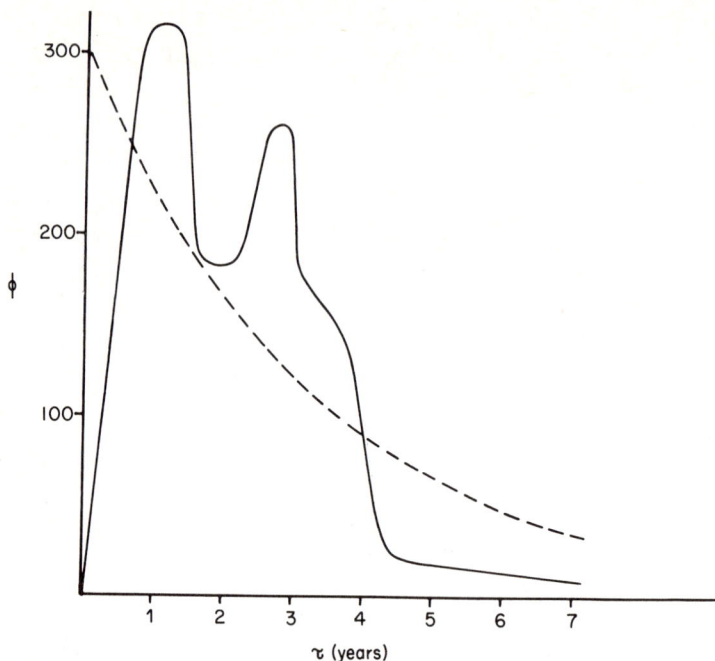

Fig. 5.1 The density distribution of transit times in the Ottawa River basin obtained by manual adjustment on data published by Brown (1961). The dashed line shows the exponential density distribution. (From Eriksson (1963), reproduced by permission from the Editor of *Tellus*).

Brown's calculation. This density distribution is expressed by $(1/\tau_0)\, e^{-\tau/\tau_0}$.

There are some unpublished data on river water tritium in Scandinavia in which a more analytical approach was used. It indicates that at most only a three parameter model of the transit time density distribution can be used with any confidence. This is in accordance with the conclusions drawn earlier on the basis of the groundwater flow pattern in basins.

The present use of tritium in groundwater investigations is mainly qualitative. Considering an analytical detection limit of 0.5 TU, the absence of tritium in a groundwater sample means that the groundwater must be of pre-bomb age. If it contains just detectable tritium, then it may be pre-bomb (and the age can be estimated) or it may be post-bomb, i.e. formed after 1952. At higher concentrations than 20 TU, the water was certainly formed after 1952. As an additional difficulty the samples containing tritium may be mixtures between very old water and very young water. A ^{14}C analysis can usually help to decide if such mixing has taken place. Tritium alone does not always give an unambiguous answer even qualitatively.

106

5.1.2   *Radioactive carbon,* ^{14}C

Libby, (Begemann and Libby, 1957) who developed the methodology for
analyses of tritium in natural waters, also investigated ^{14}C and disclosed its
various applications; in particular, the dating of ancient objects, a work
which after a long delay, rendered him a well deserved Nobel Prize. The
carbon-14-dating method became rapidly a most valuable tool in
archaeology and quaternary geology. Laboratories for this purpose which
require rather expensive and specialized techniques are now found in many
countries of the world.

^{14}C-dating can strictly speaking be done only if the atmospheric
production of ^{14}C has remained constant in the past. There may be some
periodicities in the production rate connected to the sunspot cycle but these
are usually ignored. The other requirement for dating is a knowledge of the
zero age concentration of ^{14}C in the objects to be analysed. For organic
remains, the so-called modern wood or, better, recent wood is used as a
reference. The term 'recent' implies that the reference wood was collected
before 1952 so that any bomb ^{14}C is not included. If partial decomposition of the sample has taken place, a fractionation of the isotopes
may take place. This can, however, be checked by analysing for ^{13}C as well.
If the sample has, for example, one part per thousand more ^{13}C than the
reference then it would have had two parts per thousand more ^{14}C than the
reference, if the sample was of zero age. If a sample has only 80% ^{14}C in
comparison with recent wood and its ^{13}C content is 1% higher than recent
wood, then the initial ^{14}C content is 102% of ^{14}C in recent wood. The
fraction of ^{14}C left in the sample is thus $80/102 = 0.784$. The ^{13}C based
correction hence corresponds to an age difference of about 170 years in this
case.

For dating groundwater, the inorganic fraction is used, although the
complications are considerable. When limestone is present, the carbon
dioxide in the soil air which originates from recent carbon will react with
limestones. These are usually millions of years old and consequently
completely depleted of radioactive ^{14}C. Even at an age of 100 000 years, so
little ^{14}C is left that detection by ordinary methods becomes impossible.
There are methods which may extend the ^{14}C dating back to that age but
they are not yet developed sufficiently.

In a closed system the recent carbon dioxide would be mixed with 'dead'
carbon dioxide in the proportions 1:1 so that the apparent age of even
recent water would become about 5700 years. In an open system, i.e. in the
intermediate zone, conditions are different. The root zone produces a
steady stream of recent carbon which exchanges all the time with carbon
dioxide in the intermediate zone through gaseous diffusion. This process

will remove some of the 'dead' carbon from the limestone being replaced by recent carbon dioxide. But the extent to which this removal of old carbon occurs is difficult to assess without a thorough knowledge of (a) the rate of production of recent carbon dioxide; (b) the water content of the intermediate zone and its effective porosity; (c) the position of the limestone which is dissolved by the carbon dioxide; (d) the rate of recharge of water, and possibly a few things more. This is almost impossible as it is necessary to know the date at which the groundwater under investigation was formed. Without any further knowledge than the ^{14}C data, all that can be said is that the apparent age of groundwater taken at some point downstream of the recharge area should be corrected by subtracting a figure of between 0 and 5700 years. The extremes, 0 and 5700 are perhaps easier to decide upon than anything between. Samples of recent groundwater from recharge areas can be used as a guide but with lessening conviction as the groundwater age increases, because conditions in the recharge areas at the time of recharge may have been very different. However, lacking other information, this approach could be used.

Additional isotopic information may be obtained from ^{13}C analyses of the samples. ^{13}C is usually expressed in parts per thousand deviations from a well-known standard, being at present a belemnite from the so-called Pee Dee formation in the US. If the ratio $^{13}C:^{12}C$ is R then the parts per thousand deviation, $\delta^{13}C$ is given as

$$\delta^{13}C = 1000(R_s - R_r)/R_r$$

where R_s is the sample ratio and R_r the reference ratio. The physical chemistry of ^{13}C and ^{12}C was surveyed by Deines *et al.* (1974). The equilibrium coefficients are temperature dependent. Since they are all close to unity, one can, with a slight approximation write

$$\delta^{13}C_{CO_2(aq)} = 1000(K_0 - 1) + \delta^{13}C_{CO_2(g)}$$
$$\delta^{13}C_{HCO_3^-} = 1000(K_1 - 1) + \delta^{13}C_{CO_2(g)}$$
$$\delta^{13}C_{CO_3^{2-}} = 1000(K_2 - 1) + \delta^{13}C_{CO_2(g)}$$
$$\delta^{13}C_{CaCO_3} = 1000(K_3 - 1) + \delta^{13}C_{CO_2(g)}$$

K_0, K_1, K_2, and K_3 are given in Table 5.1. Considering a temperature of 5°C one gets the following relations

$$\delta^{13}C_{CO_2(aq)} = -0.828 + \delta^{13}C_{CO_2(g)}$$
$$\delta^{13}C_{HCO_3^-} = 9.727 + \delta^{13}C_{CO_2(g)}$$
$$\delta^{13}C_{CO_3^{2-}} = 7.888 + \delta^{13}C_{CO_2(g)}$$
$$\delta^{13}C_{CaCO_3} = 11.840 + \delta^{13}C_{CO_2(g)}$$

Table 5.1 Equilibrium coefficients for the ^{13}C species in water solution at selected temperatures (taken from Deines *et al.* (1974))

Temperature (°C)	0	5	10	15	20	25
K_0	0.999175	0.999172	0.999169	0.999166	0.999164	0.999161
K_1	1.010258	1.009727	1.009225	1.008748	1.008296	1.007866
K_2	1.008308	1.007888	1.007491	1.007114	1.006757	1.006417
K_3	1.012468	1.011890	1.011342	1.010823	1.010331	1.009964

It is seen that there is practically no fractionation of the carbon isotopes in carbon dioxide between the gas phase and liquid phase.

Additional information on the initial conditions of carbon-14 during recharge of groundwater is obtained through the chemistry of water. When the partial pressure of carbon dioxide is low and only slightly alkaline but the pH is high, then weathering must have taken place in a system closed with respect to carbon dioxide. Under these conditions the $\delta^{13}C$ should be close to that of the $\delta^{13}C$ in plants, i.e. $-25‰$. No correction of initial ^{14}C concentration is necessary.

Under similar hydrochemical conditions but with evidence of calcite dissolution, the working hypothesis should be a closed system with dilution of initial recent carbon by 'dead' carbon. If the ^{13}C of the limestone is known, then the degree of mixing between recent and fossil carbon can be estimated from the ^{13}C of the inorganic carbon. If this is, say, $-12.5‰$ and the limestone is $0‰$, then the initial ^{14}C must have been 50% diluted by fossil carbon. If the ^{13}C is found to be $-20‰$, then the initial ^{14}C activity must have been only 80% of recent carbon.

Under the conditions described, the age assessment is a fairly simple procedure. The real problem may be to get good indications that the system was closed at the time when recharge took place. The chemistry of the water seems to be the only guide to this. In such cases, it may be well-worth simulating the chemical development to test the closed system hypothesis.

The open system is in many respects far more complicated than the closed one when interpreting carbon isotope data. The criteria of an open system are high partial pressure of carbon dioxide and in particular a high alkalinity provided the mineral assembly in the water-bearing rock permits it. If the water is very old and there are primary minerals present, the partial pressure of carbon dioxide may have dropped from the original open system level. Again, simulation of the chemical evolution of the water can be of considerable help in assessing the conditions during groundwater recharge at the time of recharge.

In the case of an open system and no carbonate minerals in the water-bearing formation, the source of carbon dioxide will have the $\delta^{13}C$ composition -25%. As bicarbonate is formed during the weathering processes the $\delta^{13}C$ increases because of the bicarbonate fraction of dissolved inorganic carbon. At $5°C$ the bicarbonate, which will be dominating, is enriched by 9.7% relative to gaseous carbon dioxide. If bicarbonate makes up 90% of the total inorganic carbon, the rest being dissolved carbon dioxide, then the $\delta^{13}C$ of the water sample would be $0.9(-15.3)+0.1(-25.8)=-16.4\%$, thus 8.6% higher than the source of carbon dioxide. As a consequence of this, the initial ^{14}C would have been enriched by 17.2%, i.e. 1.7%. The initial ^{14}C concentration would have been 101.7% of 'recent' carbon in plants. Here a correction appears which does not apply to a closed system. On the other hand one would not expect a ^{13}C of -16.4% in a closed system unless dissolution of marine calcite had taken place.

In an open system where dissolution of marine calcite takes places bicarbonate is formed which will initially have a ^{13}C composition of -12.5%. If this is the only source of bicarbonate then the situation is fairly uncomplicated. Exchange with the gas phase would bring the bicarbonate $\delta^{13}C$ towards its equilibrium value, which is at $5°C$, as already stated, -15.3%. And, considering the total inorganic dissolved carbon this would be -16.4% at equilibrium under the same conditions as referred to before. At this stage, there would be no dilution by fossil carbon whereas at -12.5% there would be 50% dilution. The range in ^{13}C values for calculating the fraction of fossil carbon in the sample is obviously not very large; about 4% for an age span of nearly 6000 years. The range is, however temperature dependent and increases with temperature to a certain extent. At $25°C$ the bicarbonate enrichment is 7.9% so the bicarbonate would be -17.1% and the total dissolved inorganic carbon would be close to -18%. The range in this case is about 5.5%.

If part of the bicarbonate stems from weathering of primary minerals, there would be less dilution by fossil carbon initially. The correction for the dilution present would be the same, based on the ^{13}C chemical composition and temperature during recharge. The last item may be difficult to assess when the water is very old since climate may have been different.

The actual correction for fossil carbon dilution is simple. If a is the equilibrium value of $\delta^{13}C$ of total inorganic dissolved carbon and the measured value is x then the initial ^{14}C activity is expressed by

$$A_0 = 100 - 50(a-x)/(a+12.5) + 0.2(x+25)$$

in percentage. The last term is the correction due to fractionation, at most amounting to 2%. The formula assumes a $\delta^{13}C$ for calcite of 0‰ and a $\delta^{13}C$ for gaseous carbon dioxide of $-25‰$.

One possible source of error in limestone areas is the exchange of carbon between the gas phase and the crystalline phase. Such an effect would mean in practice that the carbon reservoir of the inorganic carbon flow system is proportionately larger than the water reservoir in the water flow system. The turnover time of dissolved inorganic carbon therefore becomes larger than that for water. ^{14}C-dating, if properly assessed, will in such cases always give too large an age. There is no indirect way to establish the effect of exchange because it does not affect ^{13}C. The only possibility is experimental studies on rock samples.

^{14}C-dating in hard rock areas (except for old limestones) is beset with a number of problems. When calcium bicarbonate is derived from the weathering of primary minerals, all the carbon is biological and there would be no fundamental uncertainty of the type encountered in limestone rocks. In the intermediate zone the gaseous phase ^{13}C is set, i.e. $-\delta^{13}C$ will be $-25‰$. There will be a fractionation when bicarbonate is formed, depleting the gaseous phase of ^{13}C, but gaseous exchange with the root zone would decrease this depletion. If gaseous exchange is effective then the $\delta^{13}C$ of bicarbonate at $+5°C$ would be $-25+9.7=-15.3$ parts per thousand. If gaseous exchange is almost inhibited the bicarbonate would have a $\delta^{13}C$ approaching -25. There is, thus, a range from $-25‰$ to about $-15‰$ in the $\delta^{13}C$ of bicarbonate in the water that leaves the intermediate zone flowing into the saturated zone. The extreme $-25‰$ applies actually to a closed system but the ^{14}C age is independent of this as long as the usual correction for fractionation is applied to the analysis. It should be noted that release of calcium through weathering is not essential in this case; the bicarbonate may be accompanied by magnesium or potassium or sodium or a mixture of these. The important part is the inorganic carbon. It should also be noted that there is no 'dead' carbon to worry about in these rocks, at least not in the intermediate zone. If nothing out of the ordinary is taking place, the dating procedure should be quite safe and no other correction to bring the sample age to the biological level other than that based on ^{13}C, is needed.

In the saturated zone the mineral transformation in the now closed system will increase the bicarbonate concentration but as long as no external addition of inorganic carbon takes place there are no complications.

Precipitation of calcium carbonate may take place since continuing mineral transformation produces hydroxyl ions. There will therefore be a decrease in inorganic carbon but fractionation effects can always be at least roughly accounted for using the water chemistry data. Age estimates based on ^{14}C should therefore be safe also from this point of view.

If dissolution of previously precipitated calcite takes place then some old carbon will be added to the dissolved total inorganic carbon and this will upset the age determination. The δ^{13}C of calcite formed in fractures at deeper levels is in general now known and can be anything from $-13‰$ to $0‰$ depending upon the conditions under which most of the weathering took place – in the intermediate or the saturated zones. There seems to be no way to estimate the 'dead' carbon fraction dissolved in groundwater in such cases. It could be as high as 50% and as low as 0. Nor can one get any help from the chemistry of the water.

Summarizing the discussion on ^{14}C-dating of groundwater it can be stated:

(a) When the aquifer is a marine limestone it is always possible, using information on the ^{13}C and the water chemistry, to correct for dilution of the inorganic carbon by 'dead' inorganic carbon. If the limestone has a large specific surface there may, however, be carbonate ion exchange between the groundwater and the limestone which dilutes the ^{14}C to a degree which is difficult to assess.

(b) In igneous rocks there should be no dilution by 'dead' carbon unless the hydrochemistry of the percolating water has changed in the past in that previously precipitated calcite is dissolved, diluting the water solution with old carbon. ^{13}C data are of no help in identifying such a process.

In all cases the normal correction of ^{14}C data for fractionation effects is done on the basis of ^{13}C. If, for instance, the δ^{13}C of the sample is found to be -20 parts per thousand this means an enrichment of $-20-(-25)=5‰$ of ^{13}C and, consequently, 10‰ of ^{14}C. Hence, to make the ^{14}C comparable to recent wood (used as standard) the fraction ^{14}C should be decreased by 0.01 before it is converted into age by the equation

$$T = -8267 \ln (f_{14}) \text{ years}$$

where f_{14} is the corrected fraction of ^{14}C in the sample.

5.1.3 Radioactive chlorine, ^{36}Cl

This isotope has such a long half-life that it is suitable for dating very old groundwater. Whenever chloride in an aquifer is of recent origin, i.e. is not dissolved from old salt deposits but is most likely air borne from ocean areas, ^{36}Cl can be used. It was applied recently in a study of groundwaters of the Great Artesian Basin in Australia with considerable success. The analytical capacity for ^{36}Cl will no doubt increase in time (at present a particle accelerator is required which is quite expensive); the interpretation of the data, however, is quite simple.

5.2 Stable isotopes of hydrogen and oxygen

There are two isotopes in water which are of considerable interest in hydrology. One is deuterium, ^2H or D which has a natural abundance of 0.0151% and the other is ^{18}O, with a natural abundance of 0.2%. The molecular species of interest are hence $H_2{}^{16}O$, $H_2{}^{18}O$, and $HD^{16}O$, other combinations being rare. There are small differences in the physical properties of the species. $H_2{}^{18}O$ has a somewhat lower water vapour pressure than $H_2{}^{16}O$ and the same is true for $HD^{16}O$. During evaporation and condensation these slight differences lead to fractionations of the isotope. Consider equilibrium between the vapour and the liquid phase. The vapour phase has a smaller fraction of ^{18}O or deuterium than the liquid phase. Although the physics of fractionation during condensation is simple and entirely predictable for a system where the condensed water is immediately removed, the fractionation process in the atmosphere is complicated through turbulent mixing and isotopic exchange between water vapour and liquid water in oceans, lakes and streams. The fractionation of water is for this reason always assessed by analysis although the general pattern is qualitatively what one would expect.

At present there is a considerable knowledge on $\delta^{18}O$ and δD in rain and snow. A global network of sampling stations for monthly samples was set up in 1962 and has operated from then onwards with modifications in station distribution as required by the data. The network is a joint WMO–IAEA program and analysis is carried out at the IAEA headquarters.

The levels and seasonal variation are fairly easy to assess at almost all places on the continents. The hydrologically important features of stable isotopes in precipitation can be summarized as follows:

(a) The δ-values decrease towards high latitudes.
(b) The δ-values decrease inland.
(c) The δ-values decrease with altitude.
(d) The δ-values show a seasonal variation best developed at temperate latitudes. Summer values are higher than winter values.
(e) There is a considerable amount of random fluctuations superimposed on the regular time averaged features.

These conclusions are in fair accord with the physics of fractionation of water vapour on condensation and removal of the condensed fraction. There is another effect which is somewhat puzzling, the 'amount' effect, termed so by Dansgaard (1964) who noted that the δ-values were much lower during months with high rainfall than during months with low rainfall, in tropical regions, for example. This effect is important in semi-arid areas since groundwater recharge will usually take place only after heavy rainfalls. Low δ-values are found, for example, in the so-called Nubian Sandstone Formation in North-East Africa, and have been interpreted as being due to recharge during a pluvial period when this area had a near temperate climate.

The stable isotopes are good tracers for judging the origin of groundwater in hilly areas because of the altitude effect. The seasonal variation has been used for studies of the run-off process. The applications in hydrology will be discussed in Chapter 6.

The analytical techniques for stable isotopes of oxygen and hydrogen are well developed. There are mass spectrometers specifically dedicated to ^{18}O and deuterium analyses with a high degree of automation. The precision of ^{18}O analyses is close to 0.1 parts per thousand which can be compared to the seasonal variation of 10–20 parts per thousand. The sample size required is at most 20 ml. The technique for oxygen-18 analysis is to equilibrate the sample with carbon dioxide and make the determination on the carbon dioxide. For deuterium very small samples are required. In the procedure the sample is converted into hydrogen gas which is put into the mass spectrometer.

^{18}O and deuterium in natural waters are closely related and as a general rule

$$\delta D = 8 \ \delta^{18}O + 10$$

with usually small deviations. Consequently, there is often no point in analysing for both isotopes. If $\delta^{18}O$ is assessed, very little additional information is obtained from δD except in cases where water in lakes is subject to evaporation. Then ^{18}O and deuterium behave differently, a kinetic fractionation effect increasing the fractionation of ^{18}O much more than it does for deuterium. In this case, the two together can provide somewhat more information than only one.

References

Data from the WMO–IAEA global network of sampling stations for environmental isotopes are published regularly in the Technical Report Series of the International Atomic Energy Agency in Vienna. The following reports covering the years 1953 to 1975 have been published. Nos. 96, 117, 120, 147, 165 and 192.

Begemann, F. and Libby, W. F. (1957) Continental water balance, groundwater inventory and storage times, surface water mixing rates, and world wide circulation patterns from cosmic ray and bomb tritium, *Geochim. et Cosmochim. Acta*, **12**, 277–96.

Brown, R. M. (1961) Hydrology of tritium in the Ottawa valley, *Geochim. et Cosmochim. Acta*, **21**, 199–216.

Dansgaard, W. (1964) Stable isotopes in precipitation, *Tellus*, **16**, 436–68.

Deines, P., Langmuir, D. and Harmon, R. S. (1974) Stable carbon isotope ratios and the existence of a gas phase in the evaluation of carbonate ground waters, *Geochim. et Cosmochim. Acta*, **38**, 1147–164.

Eriksson, E. (1963) Atmospheric tritium as a tool for the study of certain hydrologic aspects of river basins, *Tellus*, **15**, 303–08.

Fritz, P. and Fontes, J. C. (eds) (1980) *Handbook of Environmental Isotope Geochemistry*, Elsevier, Amsterdam.

Gat, J. R. and Gonfiantini, R. (eds) (1981) *Stable Isotope Hydrology. Deuterium and oxygen-18 in the water cycle*, IAEA, Vienna.

IAEA (1983) *Guidebook on Nuclear Techniques in Hydrology*, IAEA, Vienna.

Further reading

Eriksson, E. (1958) The possible use of tritium for estimating groundwater storage, *Tellus*, **10**, 472–78.

Lal, D. and Peters, B. (1962) Cosmic ray produced isotopes and their application to problems in geophysics. Progress in Elementary Particle and Cosmic Ray Physics. (J. G. Wilson and S. A. Wouthuysen, eds.), vol. 6, North-Holland Publishing Co., Amsterdam.

6

Applications of hydrochemistry and environmental isotopes

Application of hydrochemistry and environmental isotopes can be envisaged in two ways. Applications may pertain to hydrological problems; how can one use hydrochemical information for hydrological purposes? Establishing the origin of groundwater and estimating its recharge rate are common examples of the type of problems concerning groundwater.

The other way to look at application is from the point of view of chemistry. What is the budget of chemical constituents in a basin and how is this budget affected by man's activity? Water quality is a consumer term. Is the water suitable for a particular purpose and how will it be affected by extraction or by addition of other substances? Also such problems are applied problems in the sense that they require knowledge and understanding of hydrochemical processes.

6.1 Hydrochemical monitoring

Monitoring is part of an environmental control activity where a number of indicators, i.e. environmental properties, are being watched to see if any systematic environmental changes are taking place in a given time period. Monitoring of hydrochemistry is usually the task of a government agency that is responsible for protecting the environment against undue changes. A lot of money is spent on this and therefore, the question as to how monitoring should be carried out is of great importance in order for it to be as effective as possible.

Monitoring usually entails sampling the environmental variables under study at intervals so that a sequence of data is obtained. The sampling should be done so that a maximum amount of information is obtained for the means allocated to the work. If it is found that in order to obtain the desired information, modifications of the programme are required, the allocation may have to be revised. This is no doubt a sound strategy since it will keep up the interest in the monitoring work, preventing it becoming just a dull routine.

6.1.1 *Selection of sampling intervals*

Continuous monitoring of an environmental variable will give a time series of observations which contains a broad spectrum of information. Thus one can, for instance, note the fluctuations of various time scales – hourly, daily, weekly, monthly and yearly variations. Some of these will be strictly periodic like diurnal and seasonal variations. They are therefore regarded as deterministic in the sense that they are predictable. In a monitoring programme, seasonal variation is of interest since a change in amplitude could be an indication of undue environmental change. The diurnal variation in water systems is not of much interest since it is often weak. For that reason, it is usually ignored. There may be cases when short-term variations are of interest, for example, in release of waste from factories, but such monitoring is then part of the control system of the factory and not of a government agency. The latter is more concerned about the status of the environment in a long term sense.

Thus, disregarding these diurnal variations in environmental monitoring programmes, the variations of interest are the seasonal periodicity and the fluctuations superimposed on the seasonal variation. These fluctuations are conveniently regarded as random since they may be predictable only in a limited sense. The spectrum of the fluctuations is usually of the so-called 'red noise' type, implying that most of the variance is concentrated at the low frequency end of the spectrum. The physical reason for this is the mixing in the water systems. Water has to pass through fairly large reservoirs, such as the groundwater storage. If the transit time distribution for these reservoirs are exponential, they behave as if rapidly mixed with respect to discharged groundwater. Consider such a reservoir with a turnover time τ, input x_t and output y_t at time t. The balance then requires that

$$y_t - y_{t-1} = -(\Delta t/\tau)(y_t + y_{t+1})/2 + (x_t + x_{t-1})/(2\tau)$$

when using finite intervals of length Δt. The relation can be rewritten as

$$y_t = (2\tau - \Delta t)/(2\tau + \Delta t)y_{t-1} + (x_t + x_{t-1})/(2\tau + \Delta t)$$

Writing $r = (2\tau - \Delta t)/(2\tau + \Delta t)$ and $z_t = (x_t + x_{t-1})/(2\tau + \Delta t)$ then

$$y_t = ry_{t-1} + z_t$$

which is an autoregression process. When z is uncorrelated – as random inputs in hydrological systems frequently are – the equation depicts a so-called first order Markov process. The spectrum of such a process is simple and has the features mentioned. Most of the variance is concentrated at the low frequency end of the spectrum.

In a random sequence y_t which is highly autocorrelated (i.e. r is close to unity) the information content in each observation is fairly small. If r is close to unity, a value y_{t+1} can be predicted with a high probability of being correct. In the expression for r, it is seen that

$$r = 0.78 \text{ for } \tau = 4 \, \Delta t$$
$$r = 0.60 \text{ for } \tau = 2 \, \Delta t$$
$$r = 0.33 \text{ for } \tau = \Delta t$$

Since the seasonal variation is a necessary consideration, one can hardly choose a Δt larger than 1/6 of the year. It may even be advisable to use $\Delta t = 1/12$ of a year, i.e. carry out monthly samples.

The turnover time for groundwater reservoirs in the simple exponential case is of the order of a few months (i.e. for local groundwater flow systems). With monthly samples, there will not be much information in the random part of each observation. The sampling interval could be extended to two months with a corresponding saving of cost.

Considering the groundwater reservoir further it would be a sound strategy to sample, say, twice a month the first year to assess the autocorrelation and the magnitude of the seasonal component. After that one could adjust the sampling interval to a more effective scheme.

In monitoring schemes, observations are usually made on several variables, both physical and chemical. After the first few years a correlation analysis should be made to get some idea of the interdependence of the variables. This can be done as a correlation matrix using all possible combinations of variables. For any pair of variables which are highly correlated, one should be selected for future monitoring since the other does not give any additional information unless changes are expected in the correlation coefficients due to changes in land use or atmospheric input.

Representation of correlation matrices can be done by cluster description as used earlier for atmospheric chemistry data. Another example from Grip

(1982) is shown in Fig. 6.1 from an investigation on the hydrochemistry in the run-off from small forested catchments in central Sweden. In this particular case, the constituents formed two well distinguishable groups, one geochemical and the other biochemical.

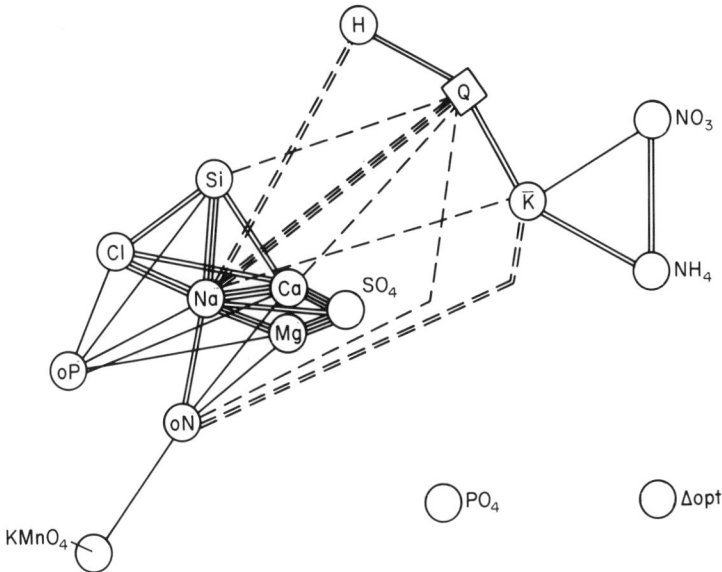

Fig. 6.1 Cluster description of associations between hydrochemical variables in run-off from small catchments (from Grip (1982)).

6.1.2 *Selection of sampling points*

Sampling can be made at more than one point in a particular area. For streams, one point at the outlet of the basin is usually sufficient but when groundwater is to be monitored several points may be needed. Considering the extreme transit time distributions possible in an aquifer, the sampling scheme should attempt to represent a wide range of transit times. The choice of sampling points must be made on the basis of topography and geology. Samples should represent points in the aquifer as closely as possible; this may call for installation of packers in boreholes or other arrangements. In an aquifer, there should be at least one point representing recent recharge of groundwater and then two or more points near discharge areas representing the local, the intermediate and the regional groundwater flow systems. The installation cost of such schemes may be high but it seems to be the only meaningful arrangement for monitoring groundwater chemistry.

In a groundwater monitoring scheme of this type, the sampling intervals may vary at the different points. The recent recharge sampling point should be sampled every month for the first year, after which a suitable interval is selected, perhaps every three months. In the discharge area, the local flow system should be sampled at least four times a year, perhaps even more frequently in the first year. For the intermediate and regional flow systems, however, one sample per year should be sufficient.

Monitoring schemes in lakes are mostly concerned with biology in a more general sense and are therefore outside the scope of this book. Chemistry is, of course, part of the monitoring scheme. For this purpose, two representative sampling points would be sufficient, one for surface water and the other for deeper water. Temperature profiles would be needed prior to sampling to delineate water above and below the thermocline whenever such a feature is present. The sampling frequency at the surface water point should be adjusted to fit the biological programme. For deeper water, sampling twice a year would be sufficient if the turnover time of the water in the lake is of the order of a year or more; if less than this, more frequent sampling is required.

6.1.3   Data analysis

Data analysis has two purposes. One is to see if modifications in the programmes are needed, such as changes in sampling intervals or changes in the number of variables to monitor. This part has been discussed to some extent where it was pointed out that correlation analysis is a good method to use in this respect. The other, and perhaps the essential purpose of data analysis is to detect trends of such a nature that they can be related to land management or external influences, like noticeable changes in the deposition of airborne chemical substances. Whatever the purpose of the data analysis, a suitable procedure should be followed.

The components of a time sequence are the mean, deterministic periodicities, trend and a random process. The random process can usually be described by its autocorrelation structure and the residual variance. The last item represents a sequence of uncorrelated random variables. In the absence of trend or when it is removed the sequence is assumed to be weakly stationary in the sense that the mean and variances are time independent. If present, periodic variation of the variance can be eliminated by simple transformation of the variables. Periodic variation in the autocovariance cannot be eliminated in this way.

An example of the analysis of sequences of monthly data will be given in this section. The variables are pH, alkalinity, electrical conductivity, and concentrations of sulphate, chloride, sodium, potassium, magnesium, and calcium in two streams in the so-called representative basin Kassjöån in Sweden, investigated during the International Hydrological Decade. This basin is situated in the middle of Sweden and represents a fairly typical forested area with a topography indicating strongly fractured precambrian rocks.

The two stations chosen for sampling are at Kroksillret and Norrsjön. Kroksillret represents drainage from the northern part of the basin, an area of 76.7 km² and Norrsjön the most distant part of this sub-basin with a catchment area of 15.3 km². The sampling point at Kroksillret is situated downstream of a fairly large lake, Lake Tivsjön, 3.9 km² in area with a mean depth of 11 m. Upstream of Norrsjön there is a small lake, the area of which is 0.3 km². There are five years of complete data from 1970 to 1974 which will be used in the present context (Falkenmark 1979). The final set of data is found in Appendix C.

Check of the data
The first step is to look through the tables for obvious anomalies. At the Kroksillret station the November 1970 value of alkalinity was about four times higher than usual; similarly for calcium and also the electrical conductivity. These variables were adjusted to the values of previous months. The next month, December 1970, showed a remarkably low calcium concentration which also upsets the charge balance of ions but it was accepted although it may have been in error. In situations like this too much selection of data may introduce bias which must be avoided. If a selection is needed one should use a fairly generous range of tolerance, say, three times the standard error.

For Norrsjön there were no data for February 1972. A set of data for this month was worked out from means between adjacent months and used in the final data set. As a consequence, the number of degrees of freedom in the statistical tests should be reduced by one, although in a sample of 60 this does not seem to be necessary.

Correlations and cluster description
In a set of data of the kind described the association between the variables is often of great interest, particularly for interpretation of weathering processes in the ground. For calculating correlation coefficients standard practices are used.

Correlation between different variables is conveniently described graphically in cluster diagrams, in one way or another. One type was described previously. In the present case, there are two ways of carrying out correlation. It can be done using the concentration data, in which case one can expect correlation due to changes in the general salt content of the water. There will be a dilution during snow melt and an increase in concentration due to evapotranspiration during summer. The clustering under these circumstances is shown in Figs 6.2 and 6.3 for Kroksillret and Norrsjön respectively. Not surprisingly, the covariation in concentrations is well expressed by the major ions. There are, however noteworthy differences between the two stations. The calcium concentration at Kroksillret appears to be independent of the others whereas at Norrsjön, apart from potassium, it is strongly associated with the major ions.

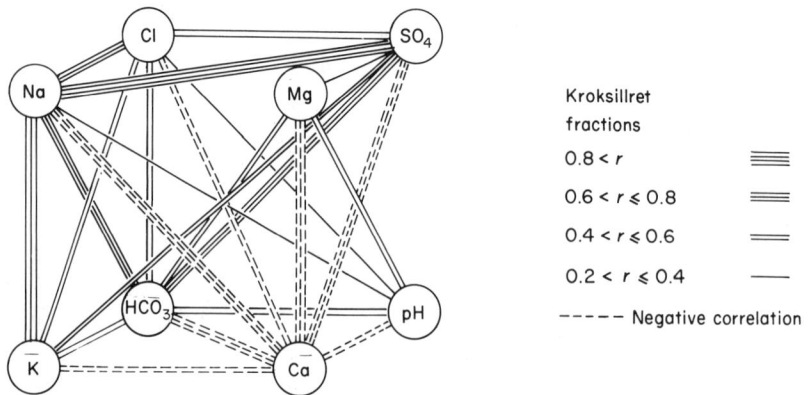

Fig. 6.2 Cluster description of correlations between concentrations at Kroksillret in the Kassjöån representative basin in Sweden.

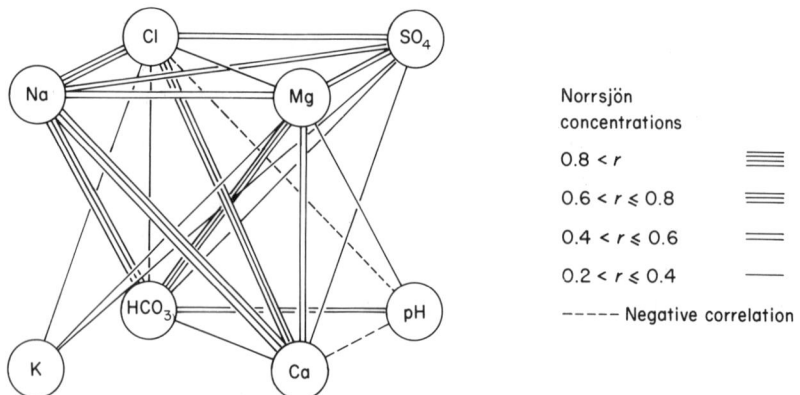

Fig. 6.3 Cluster description of correlations between concentrations at Norrsjon in the Kassjöån representative basin in Sweden.

However, potassium at Norrsjön is relatively low when compared to sodium, indicating perhaps that contribution from feldspar weathering is small. The run-off at Norrsjön belongs by and large to the local groundwater system with relatively short transit times. At Kroksillret, near the outlet of the basin, the contribution from the regional groundwater flow system should be noticeable and may therefore carry more potassium. This is, of course, speculation well-worth pursuing by further observations.

Correlation can also be done on relative concentrations, i.e. using fractions. In the present case, this is achieved by dividing the concentrations by the sum of cations expressed in equivalents. The charge balance strongly indicates the presence of ionized organic compounds – perhaps up to 30% of the sum of negative charges are made up of these. Hence, the sum of cations should be close to 50% of the total charges. If the total dissolved matter had been separately determined, it could have been used for calculating the fractions.

The resulting correlations are shown in the cluster diagrams in Figs 6.4 and 6.5 for Kroksillret and Norrsjön respectively. Considering first Fig. 6.5 it can be seen that the fractions of Ca are negatively correlated to all other variables. This must be interpreted to mean that when the fraction Ca increases all the other fractions decrease, particularly obvious for Mg. A few others are rather independent like K and Cl. Four variables, Na, Mg, HCO_3, and pH form a cluster with positive covariation in relative concentrations.

Looking at Fig. 6.4, the Kroksillret fractions, the difference between Ca and the other variables is even more pronounced than in Norrsjön. The Ca

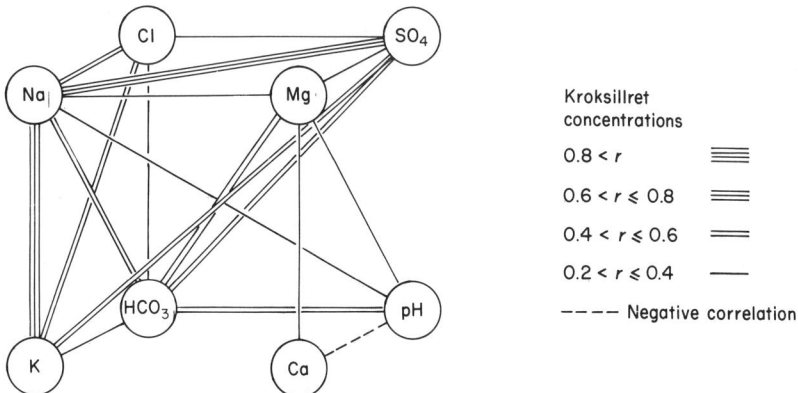

Fig. 6.4 Cluster description of correlations between fractions at Kroksillret in the Kassjöån representative basin in Sweden.

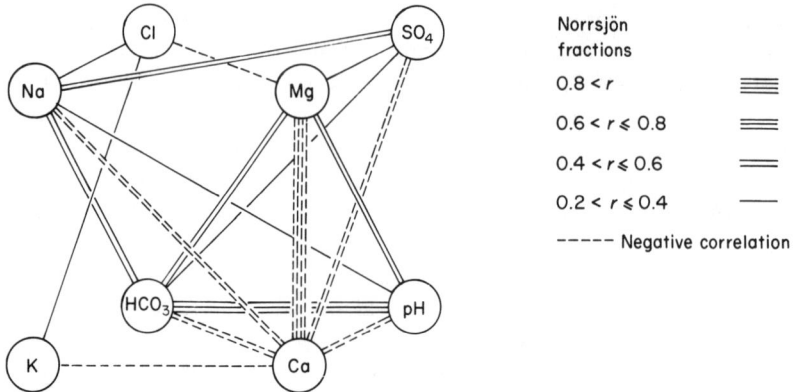

Fig. 6.5 Cluster description of correlations between fractions at Norrsjön in the Kassjöån representative basin in Sweden.

is most likely to be associated with organic acids, such as humus in solution, perhaps even as 'ion pairs', if this concept is acceptable.

Considering Fig. 6.4, a fairly well-knit cluster of ions is formed by Na, K, Cl, SO_4, HCO_3, and pH. Mg is joined to this via HCO_3, pH and, weakly, SO_4. Ca is completely outside the cluster.

Clustering can, and frequently is, depicted by so-called dendrograms, diagrams which look like a branching tree. There are several procedures available for constructing dendrograms but the one described here (Davis, 1973) is considered to be good. Dendrograms are a way of displaying the ranking and association of correlation coefficients between a set of objects although the procedures used have no statistical foundations.

To demonstrate the procedure of constructing dendrograms, the matrix of correlation coefficients between concentrations of HCO_3, SO_4, Cl, Na, K, Mg, and Ca at Norrsjön are used. Labelling these variables from A to G in the order given, the first matrix, $M1$, is shown in Table 6.1. Since the matrix is symmetrical, it always contains two identical correlation coefficients so that one can be related to a row and the other to a column.

The first step is to localize the pairs of correlation coefficients that are highest. In matrix $M1$ these are found in row A column D and row D column A and have the value 0.73. Now columns A and D can be combined into one column labeled AD simply by averaging the correlation coefficients (except where there is a 1 where AD naturally becomes 1). The matrix $M2$ shows the result. In the new matrix the column AD entry B becomes 0.40, being the average of 0.33 and 0.46 in $M1$. And entry C is the average of 0.27 and 0.72. The rows A and D are also combined in the same way so that the matrix becomes symmetrical. In matrix $M2$ the highest

Table 6.1 Successive steps in cluster analysis: demonstration using correlations of concentration data from Norrsjön*

Matrix M1	A	B	C	D	E	F	G
A	1.00	0.33	0.27	0.73	−0.04	0.62	0.34
B	0.33	1.00	0.47	0.46	0.21	0.53	0.32
C	0.27	0.47	1.00	0.72	0.37	0.30	0.63
D	0.73	0.46	0.72	1.00	0.14	0.53	0.62
E	−0.04	0.21	0.37	0.14	1.00	0.20	0.11
F	0.62	0.53	0.30	0.53	0.20	1.00	0.45
G	0.34	0.32	0.63	0.62	0.11	0.45	1.00

Matrix M2	AD	B	C	E	F	G
AD	1.00	0.40	0.50	0.05	0.58	0.48
B	0.40	1.00	0.47	0.21	0.53	0.32
C	0.50	0.47	1.00	0.37	0.30	0.63
E	0.05	0.21	0.37	1.00	0.20	0.11
F	0.58	0.53	0.30	0.20	1.00	0.45
G	0.48	0.32	0.63	0.11	0.45	1.00

Matrix M3	AD	B	CG	E	F
AD	1.00	0.40	0.49	0.05	0.58
B	0.40	1.00	0.40	0.21	0.53
CG	0.49	0.40	1.00	0.24	0.38
E	0.05	0.21	0.24	1.00	0.20
F	0.58	0.53	0.38	0.20	1.00

Matrix M4	ADF	B	CG	E
ADF	1.00	0.46	0.43	0.13
B	0.46	1.00	0.40	0.21
CG	0.43	0.40	1.00	0.24
E	0.13	0.21	0.24	1.00

*The following symbols are used for the variables:

A HCO_3	B SO_4	C Cl	D Na
E K	F Mg	G Ca	

Table 6.1   (continued)

Matrix M5	ADFB	CG	E
ADFB	1.00	0.42	0.17
CG	0.42	1.00	0.21
E	0.17	0.21	1.00

Matrix M6	ADFBCG	E
ADFBCG	1.00	0.19
E	0.19	1.00

correlation coefficient pairs are found in column C and G. Consequently, the columns C and G are combined in the same way as before into a new column CG, and the rows are modified accordingly. This results in matrix $M3$ in the table. In this, columns AD and F have the highest correlation coefficient pair and are combined into the row and column ADF. In the next step in matrix $M4$, ADF and B are combined into $ADFB$ which in matrix $M5$ is combined with CG into the final matrix $M6$, a 2 by 2 matrix. This ends the computation procedure.

The dendrogram is now constructed as shown in Fig. 6.6. The scale to the left measures the strength of 'correlations' obtained during the procedure. The first combination made, AD represents HCO_3 and Na at a correlation level of 0.73 illustrated as in the figure. Since this combines with F (i.e. Mg which combines with B, i.e. SO_4) they are arranged as in the figure. The correlation level for ADF is 0.58 and that of $ADFB$ 0.46 as seen in the table. Then comes CG as a separate branch at a correlation level of 0.63, representing Cl and Ca. This is joined to $ADFB$ at a correlation level of 0.42 (cf. $M5$ in the table). Lastly, E representing K is joined to the big cluster at a correlation level of 0.19 as seen in $M6$ of the table.

Thus, the procedure is fairly simple, at least for such a small matrix as that used here. However, except for the highest level, none of the correlation coefficients are estimated in any statistical sense. They merely serve the purpose of ranking and construction of the dendrogram.

Figure 6.6 demonstrates quite nicely the way clusters are built up. There are only two close pairs, HCO_3–Na and Cl–Ca. Comparing this diagram to the cluster description in Fig. 6.3, (disregarding pH associations) the Cl could easily have taken the role of HCO_3 with the correlation coefficient 0.72 as compared to 0.73 for the Na–HCO_3 pair. As the procedure

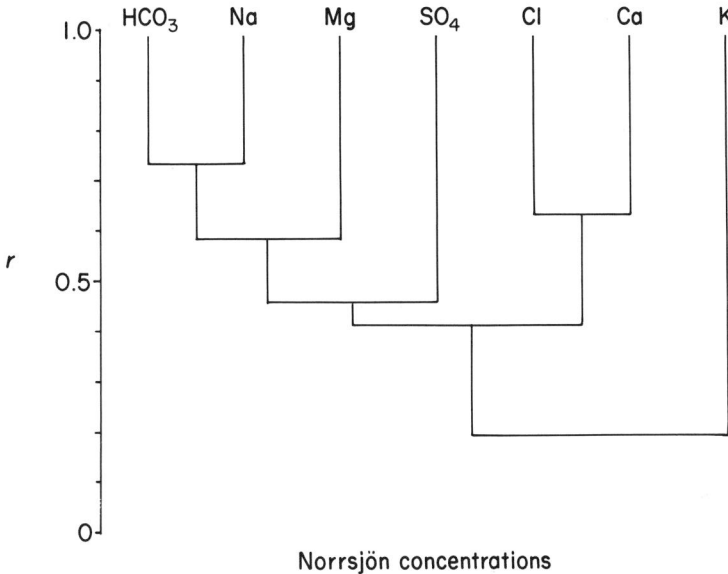

Fig. 6.6 Dendrogram showing the association of ions when represented by concentrations at the Norrsjön station in the Kassjöån representative basin.

progressed the correlation between Cl and the Na–HCO$_3$ pair was weakened. Perhaps this *is* the case, since the correlation between Cl and HCO$_3$ is rather weak. At any rate, the display of associations in a dendrogram is very clear although some reservations need to be made with respect to the interpretation. Similar dendrograms are shown in Figs 6.7, 6.8 and 6.9 for Kroksillret concentrations as well as for the fractions in both places.

Trends
There are many procedures available for detecting trends in time sequences. The most common is the use of regression analysis, usually linear regression. In this procedure a straight line is fitted to the sequence in the form

$$x_t = a + Bt$$

B being the linear trend and t the time step. There is no particular hypothesis behind this procedure so that extrapolation beyond the end of the time sequence used has no predictive value. The relation simply indicates the trend within the observation period.

Another fairly simple approach is to group the data so that a comparison can be made between early data and late data. The maximum efficiency of

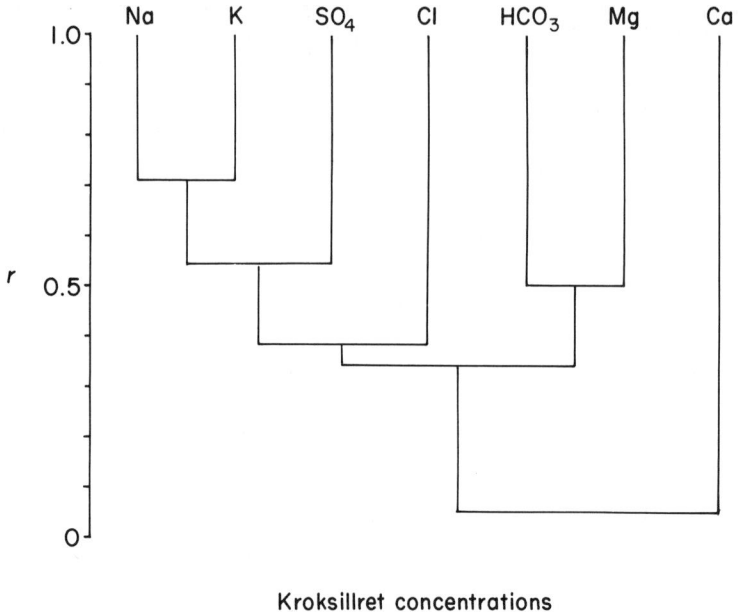

Fig. 6.7 Dendrogram showing the association of ions when represented by concentrations at the Kroksillret station in the Kassjöån basin.

this procedure is attained by comparing the sum or average of the last third of the sequence with the sum or average of the first third. The difference in averages then represents 2/3 of the time span of observations. The difference can be tested statistically using the null hypothesis as in the usual procedure for testing differences of means, provided that the time sequence is uncorrelated. Since seasonal variation is uninteresting from the point of view of long term trends, these should be based on yearly averages. In most cases such data are uncorrelated which simplifies hypothesis testing.

The time development in the two series is shown in Fig. 6.10 for all the variables considered, including the excess cations worked out from the ionic balance. This excess of cations is presumably balanced by high polymer organic acids in solution. In general, nitrate concentrations are so low that they contribute very little to the ionic balance.

Some of the variables display time developments which can be called trends, most obvious for the pH. These trends are closely related to the river discharges. From the entire experimental basin, 165 km² in area, the yearly average discharges during the five-year period were 2.0, 1.4, 1.2, 0.92 and 1.7 m³ s⁻¹. This fits very well into the pH and alkalinity trends. The electrical conductivity and cation concentrations (except for those of

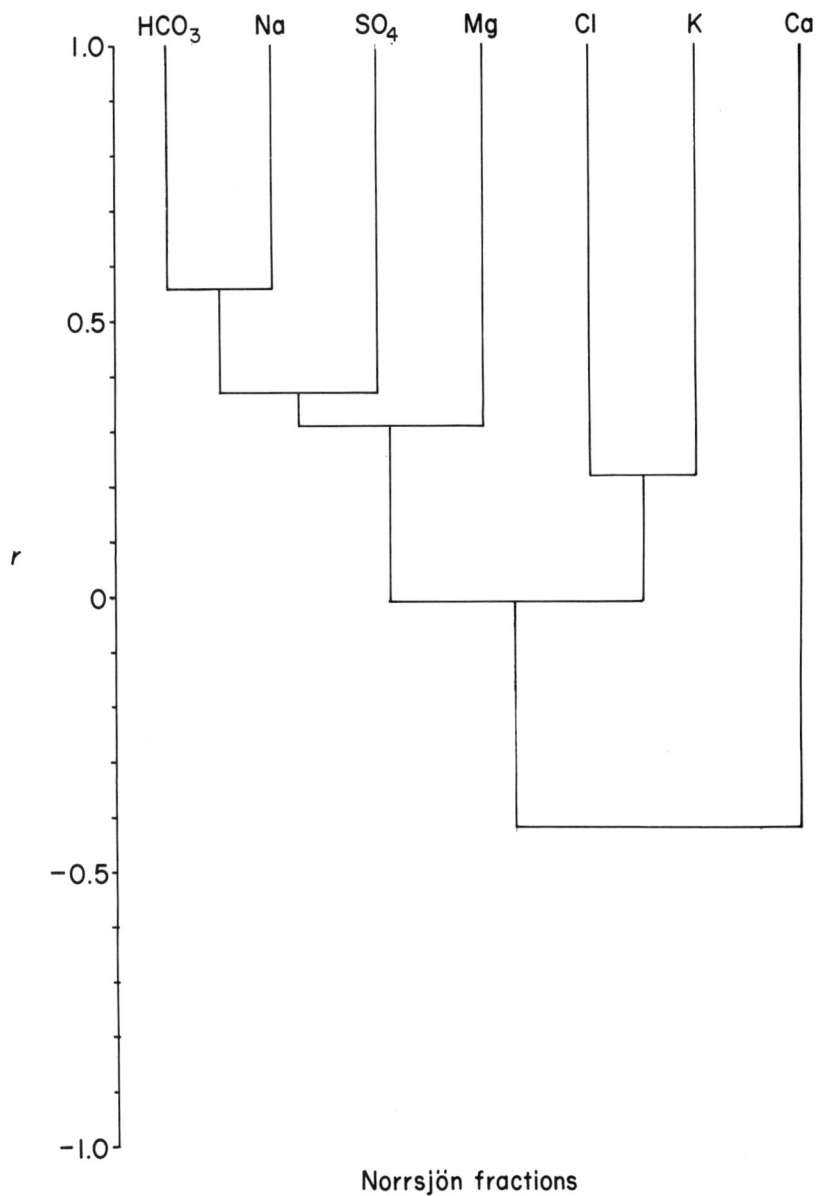

Fig. 6.8 Dendrogram showing the association of ions when represented by fractions at the Norrsjön station in the Kassjöån basin.

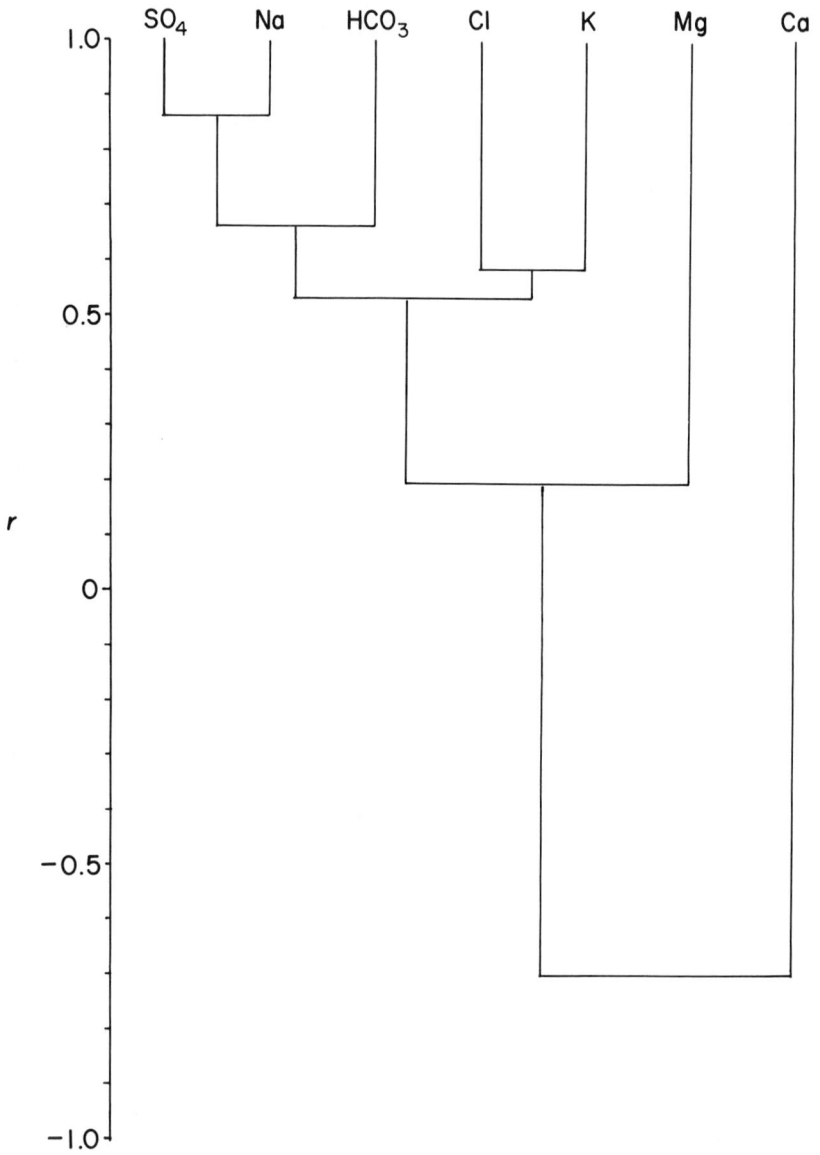

Fig. 6.9 Dendrogram showing the association of ions when represented by fractions at the Kroksillret station in the Kassjöån basin.

calcium) also follow the general trend but tend to continue afterwards presumably because of time lag in the turnover of water in the basin. Calcium behaved in a peculiar fashion during 1973, the year with the smallest discharge. The ionic balance in the same period also shows a minimum. The rainfall that summer was very low and it is possible that very little surface run-off in groundwater discharge areas took place, this run-off being mainly responsible for addition of organic matter to the water. There is, of course, the other possibility, of analytical errors in the calcium determinations, although it is unlikely in the present case.

The linear trend procedure can, of course, be applied to the data. The trend coefficient $\hat{B}$ is worked out from

$$\hat{B} = (12/(M(M+1)(M-1)) \sum_{t=1}^{M} \bar{x}_t(t-(M+1)/2)$$

the roof sign $\hat{}$ indicating an estimated coefficient. In the present case, with $M = 5$

$$\hat{B} = 0.1 \sum_{t=1}^{5} \bar{x}_t(t-3)$$

$\bar{x}_t$ being the yearly means.

The significance of $\hat{B}$ can be tested by estimating its mean error from the equation of its variance

$$\varepsilon_B^2 = 12\hat{s}^2/(M(M+12)(M-1))$$

where $\hat{s}^2$ is the unbiased variance of the sequence, i.e.

$$\hat{s}^2 = (1/(M-2)) \sum_{t=1}^{M} (\bar{x}_t' - \bar{\bar{x}})$$

$\bar{x}_t'$ being the $\bar{x}_t$ corrected for trend. In the present case this is done by

$$\bar{x}_t' = \bar{x}_t - \hat{B}(t - (M+1)/2)$$

When working out $\hat{s}^2$ for the corrected series, $M-2$ is used instead of $M-1$ indicating the loss of two degrees of freedom as opposed to 1 in the case of a single mean. This can be understood by considering the five data available. Three of them can be given any value but the fourth and fifth must be chosen such that the mean and trend of the sequence are preserved.

The significance test is simple. With 3 degrees of freedom the Student's t-value is 3.18 for a probability limit of 5%. Hence, if $|\hat{B}| > 3.18 \ \varepsilon_B$ it is considered significantly different from zero. Table 6.2 shows the linear trend coefficient $\hat{B}$ and ε_B for the nine variables at Kroksillret and Norrsjön. The column showing $B_{0.05}$ contains the lower limits of the trend coefficients that

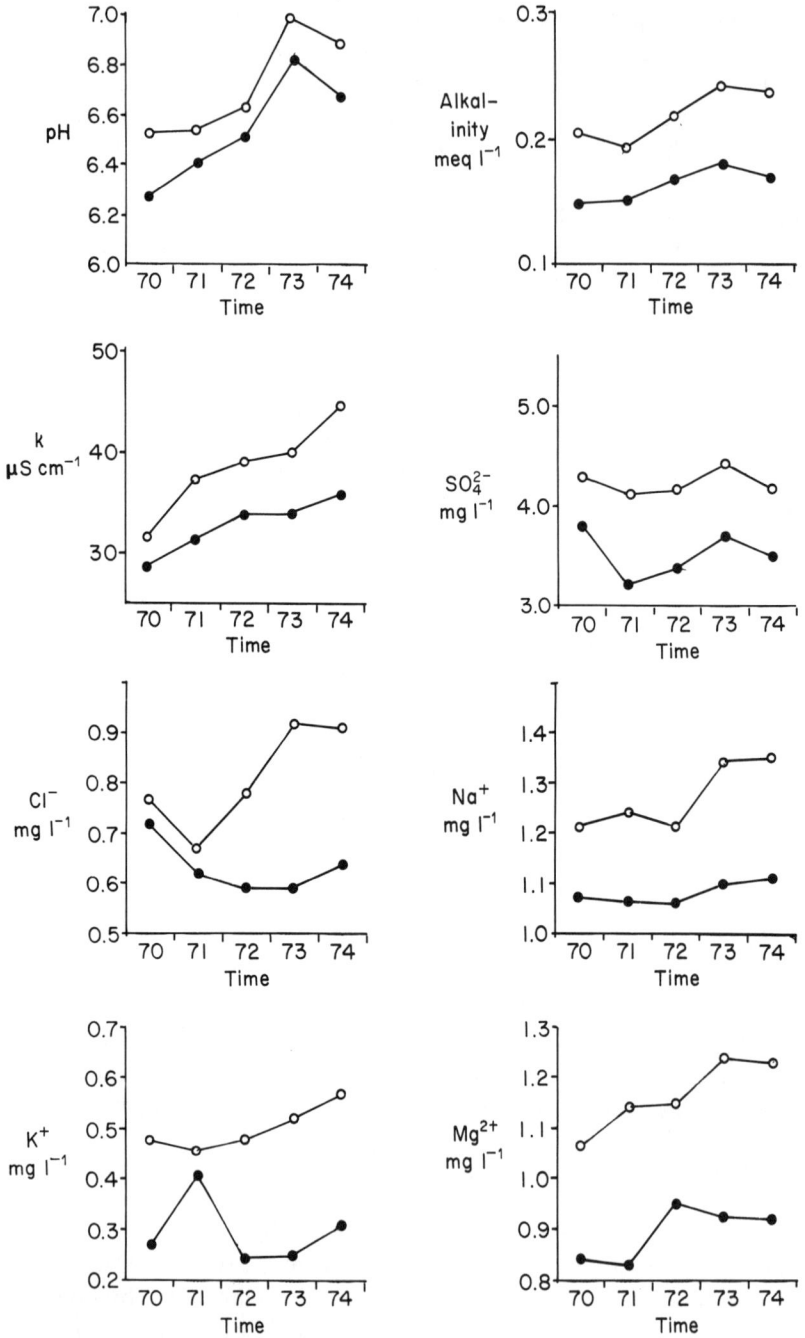

Fig. 6.10 (continued on facing page)

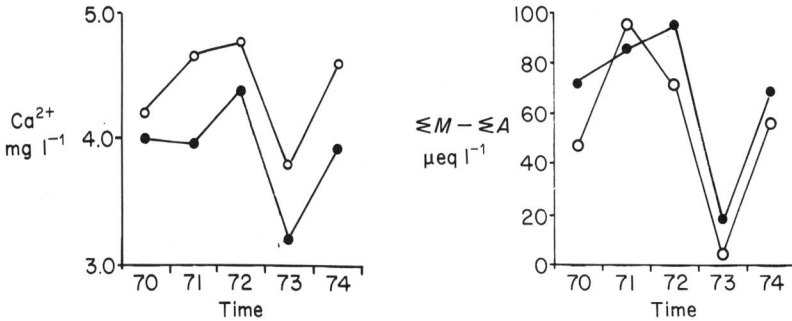

Fig. 6.10 Yearly means of the variables at Norrsjön [●] and Kroksillret [○] in the Kassjöån representative basin in Sweden.

are considered to be significant. From the table it is seen that the increase in electrical conductivity is to be regarded as significant whereas the pH and alkalinity trends are barely significant.

The trends worked out are not unexpected considering the climatic conditions during the study. Hence, one cannot extrapolate these trends for predictive purposes. They just indicate what has happened and the question about statistical significance is therefore uninteresting.

The other method which was mentioned can be applied in such a way that the sum of the last two years is compared to the sum of the first two years. In other words, a difference D is defined by

$$D = \bar{x}_4 + \bar{x}_5 - \bar{x}_1 - \bar{x}_2$$

With the null hypothesis the expected value of D is zero as long as the sequence represent a stationary random process. The estimated variance of the $\bar{x}_t$ sequence becomes

$$\hat{s}^2 = (1/3) \sum_{t=1}^{5} (\bar{x}_t - \bar{\bar{x}})$$

where $\bar{\bar{x}}$ is the average of the sequence. For the significance test, the Student's t-value, 3.18 is also used here. Since D involves four data the standard deviation of D becomes $2\hat{s}$. Thus if $|D| > 3.18(2\hat{s})$ then the null hypothesis has to be rejected. Thus, for $|D| > 3.18(2\hat{s})$ there is a significant trend in the sense that it is likely to continue in future or at least lead to a new stationary state. This trend test is rather rigorous since it does not allow any correction for trend in the time sequence when estimating the standard deviation. Table 6.2 also shows D and its standard deviation and the lower limit for 5% probability, $D_{0.05}$. It is seen that there is no significant trend in the time sequences according to this test, which thus

Table 6.2 Linear least square trends, $\hat{B}$, their mean errors, ε_B and significance limits $B_{0.05}$ in comparison to trends based on the assumption of a stationary time sequence.*

Property	$\hat{B}$	ε_B	$B_{0.05}$	D	ε_D	$D_{0.05}$
I Kroksillret						
pH	0.114	0.037	0.118	0.79	0.41	1.30
Alkalinity (meq l^{-1})	0.011	0.004	0.013	0.079	0.041	0.13
Electrical conductivity (μS cm^{-1})	2.80	0.471	1.50	15.2	9.2	29.2
Sulphate (mg l^{-1})	0.007	0.047	0.149	0.2	0.26	0.82
Chloride (mg l^{-1})	0.054	0.023	0.073	0.39	0.21	0.66
Sodium (mg l^{-1})	0.038	0.013	0.041	0.24	0.14	0.45
Potassium (mg l^{-1})	0.024	0.009	0.029	0.15	0.09	0.29
Magnesium (mg l^{-1})	0.042	0.009	0.029	0.26	0.14	0.45
Calcium (mg l^{-1})	−0.007	0.148	0.471	−0.48	0.81	2.57
II Norrsjön						
pH	0.117	0.035	0.111	0.8	0.42	1.34
Alkalinity (meq l^{-1})	0.007	0.002	0.008	0.047	0.025	0.08
Electrical conductivity (μS cm^{-1})	1.72	0.264	0.84	9.9	5.7	18.1
Sulphate (mg l^{-1})	−0.021	0.085	0.27	0.13	0.47	1.50
Chloride (mg l^{-1})	−0.021	0.016	0.05	−0.11	0.11	0.35
Sodium (mg l^{-1})	0.012	0.005	0.016	0.08	0.05	0.16
Potassium (mg l^{-1})	−0.009	0.025	0.080	−0.12	0.14	0.45
Magnesium (mg l^{-1})	0.025	0.014	0.045	0.17	0.11	0.35
Calcium (mg l^{-1})	−0.086	0.145	0.461	−0.8	0.84	2.68

*$D = \bar{x}_4 + \bar{x}_5 - \bar{x}_1 - \bar{x}_2$; ε_D standard deviation of D; $D_{0.05}$ significance limit of D at the 5% level; $B_{0.05} = 3.18\ \varepsilon_B$, $D_{0.05} = 3.18\ \varepsilon_D$.

suggests a stationary random process. The linear least square trends observed are just normal slow fluctuations typical for such processes. They depend on variation in the yearly water balance which is well modelled as a random stationary process. There are thus no detectable man-induced changes in the sequences.

In the present case there were only five data in the time sequences. For longer time sequences one should try to come as close as possible to the differences between the last and first thirds of the data. In generalized terms, a number M as close as possible to $N/3$ (N being the number of data in the sequence) is chosen. Then one proceeds by calculating

$$D = \sum_{t=N-M+1}^{N} \bar{x}_t - \sum_{t=1}^{M} \bar{x}_t$$

with the variance

$$\varepsilon_D^2 = 2M\hat{s}^2$$

$\hat{s}^2$ being computed from

$$\hat{s}^2 = (1/(N-1))\left(\sum_{t=1}^{N} \bar{x}_t^2 - (1/N)\left(\sum_{t=1}^{N} \bar{x}_t\right)^2\right)$$

Expressing D per time interval is made by $d = D/(N-M)$. The mean error of d will be $\varepsilon_D/(N-M)$. For the significance test there are $N-2$ degrees of freedom. The Student's test is used, again.

Seasonal variation
From the data the seasonal variation of each variable is obtained by averaging data for each month. Hence, one obtains the average seasonal variation. The results for the two basins are displayed in Figs 6.11 and 6.12.

The pattern of seasonal variation is a deterministic feature but in a short time sequence like the one used here the averages for each month are beset with statistical uncertainties. Looking at Fig. 6.11 neither the pH nor the alkalinity at Kroksillret appear to display any significant seasonal variation. Chloride and sodium on the other hand show systematic variations. In Norrsjön, the seasonal variation is quite pronounced with frequently a sharp drop in concentrations in May, reflecting dilution by snow melt. Figure 6.12 shows the excess cations. There is apparently no systematic seasonal variation in this.

The seasonal variation can thus be described entirely the way displayed in Figs 6.11 and 6.12. For some variables like SO_4 and Na, at Kroksillret a simple harmonic analysis can be made describing the variation during the seasons in terms of trigonometric functions – a combination of sine and cosine terms. Such a procedure does not, of course, give any additional

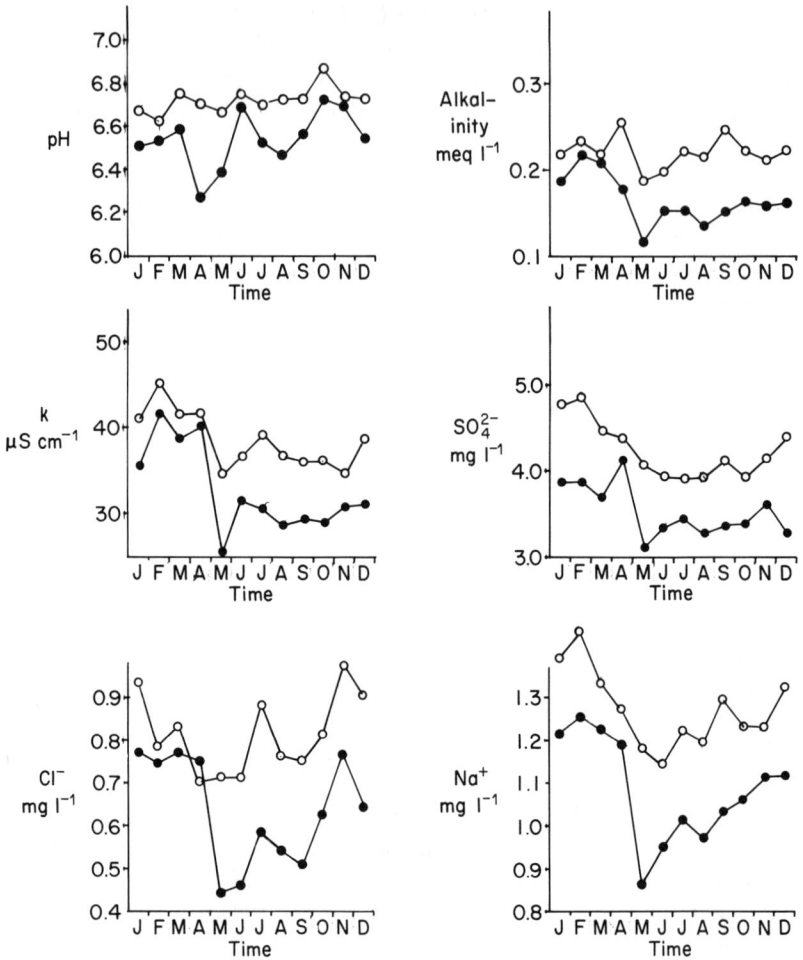

Fig. 6.11   The seasonal variation of some of the variables at Norrsjön [●] and Kroksillret [○] in the Kassjöån representative basin.

information but can be useful as a description of a model. The procedure for so-called harmonic analysis is well known and is found in standard textbooks on applied mathematics. For monthly data, a harmonic function of six sine and six cosine terms can be made to go through every point. One can, of course, use fewer components in the function and obtain a 'least squares' fit but then such drastic changes as the drop in concentrations during snow melt will not be properly described.

Harmonic analysis will be demonstrated on SO_4 and Cl in the

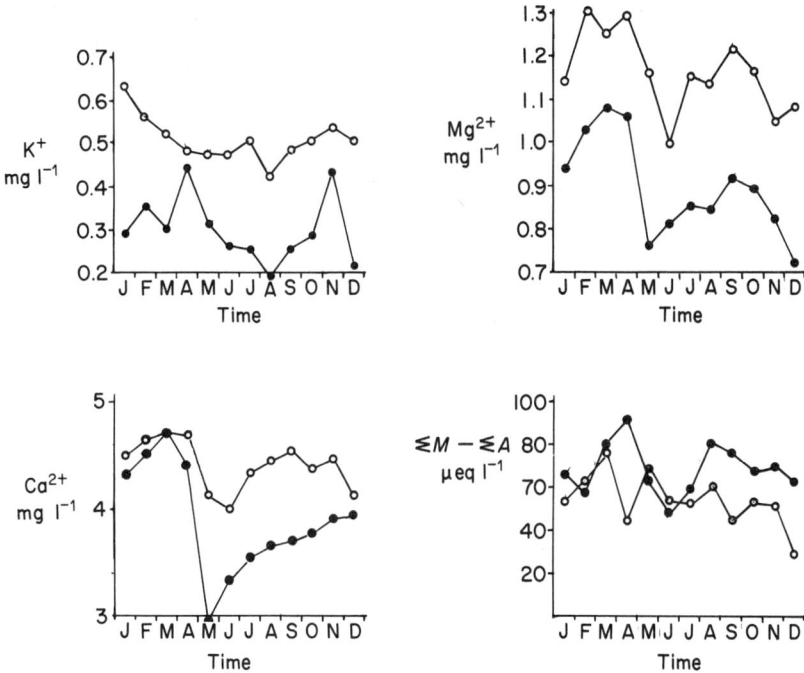

Fig. 6.12 The seasonal variation of the remaining variables (cf. Fig. 6.11) at Norrsjön [●] and Kroksillret [○] in the Kassjöån representative basin.

Kroksillret data. For this purpose it is assumed that the seasonal variation can be described by the expression

$$y_j = a_0 + a_1 \cos(2j\pi/12) + b_1 \sin(2j\pi/12) + a_2 \cos(4j\pi/12) + b_2 \sin(4j\pi/12)$$

The coefficients are worked out from

$$a_0 = (1/12) \sum_{j=1}^{12} \bar{y}_j$$

$$a_1 = (2/12) \sum_{j=1}^{12} \bar{y}_j \cos(2j\pi/12)$$

$$b_1 = (2/12) \sum_{j=1}^{12} \bar{y}_j \sin(2j\pi/12)$$

$$a_2 = (2/12) \sum_{j=1}^{12} \bar{y}_j \cos(4j\pi/12)$$

$$b_2 = (2/12) \sum_{j=1}^{12} \bar{y}_j \sin(4j\pi/12)$$

Principles and applications of hydrochemistry

$\bar{y}_j$ being the average of data from month j. The results are shown in Fig. 6.13 where the harmonic functions according to the equation given are represented by the full drawn lines, while the circles are the observed monthly means. It is seen that SO_4 is very well represented by a harmonic function which reads

$$y(t) = 4.23 + 0.25 \cos(2t\pi/12) - 0.061 \cos(4t\pi/12)$$
$$+ 0.32 \sin(2t\pi/12) + 0.13 \sin(4t\pi/12)$$

t being time in months with $t = 1$ on January 15. The fit of this type of harmonic function to the chloride data is less successful. A third harmonic

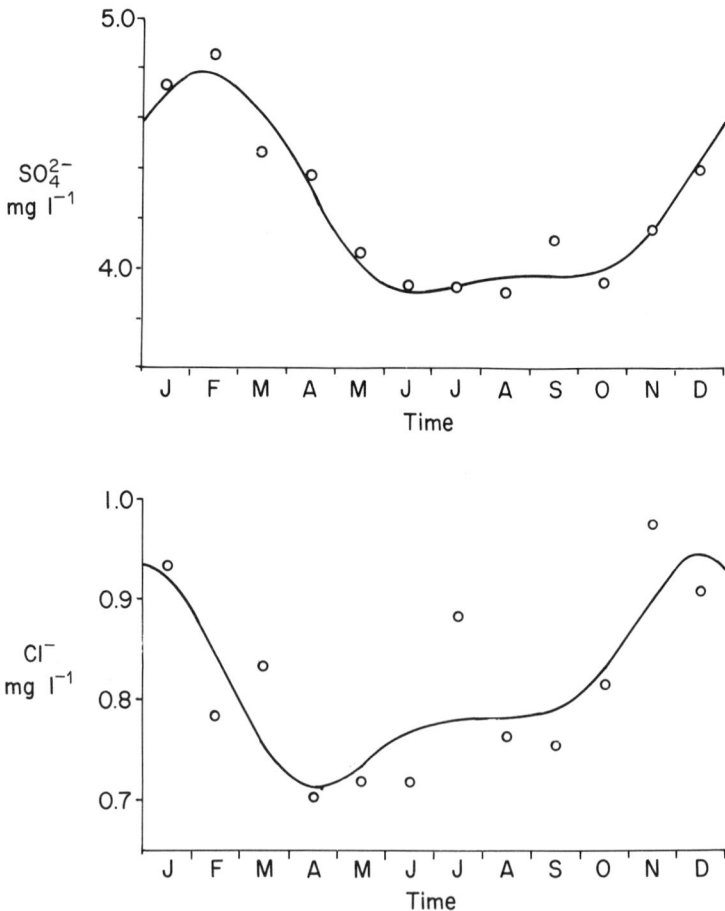

Fig. 6.13 The seasonal variation of sulphate and chloride at Kroksillret fitted to harmonic functions.

138

component with the argument $6t\pi/12$ would probably improve the fit. The present equation reads

$$y(t)=0.81+0.087\,\cos(2t\pi/12)+0.042\,\cos(4t\pi/12)$$
$$-0.017\,\sin(2t\pi/12)+0.023\,\sin(4t\pi/12)$$

The contribution of the harmonic terms to the total variance is simple to work out being half the sum of squares of the coefficients. There is a quite clear phase lag between SO_4 and Cl in the seasonal variation, SO_4 being about two months delayed. Comparing SO_4 to the other ions it appears to be in phase with Na, which is, at least partly, airborne. There are thus features in the seasonal variation of concentrations which are striking although speculation without further evidence would be unrewarding.

The random time sequence
The seasonal variation is removed from the sequence by subtracting the monthly means from the sequence. Thus, $\bar{y}_j$ being the mean value of month j is subtracted from all j months in the sequence. The remaining sequence is now free from any deterministic component. It represents a completely random and stationary process. Even the mean value is removed so that the average of the sequence is zero. One can label this sequence z_t.

The z_t sequence is now analysed for autoregression according to the model

$$z_t=r_1z_{t-1}+u_t$$

where r_1 is the correlation coefficient between adjacent data in the sequence (between z_t and z_{t-1}) and u_t is a sequence of uncorrelated random variables. Most time sequences of chemical properties will follow this simple scheme. The appropriateness can be tested by computing a set of r_k:s with increasing lags, r_1 being $\overline{z_tz_{t-1}}/\bar{z}_t^2$, r_2 being $\overline{z_tz_{t-2}}/\bar{z}_t^2$ and so on. For the simple autoregression model given the condition

$$r_k=r_1^k$$

must be fulfilled within the limits of the estimation uncertainties of the r values. There may be cases when a second order autoregression is needed, i.e. the random sequence fits

$$z_t=\lambda_1z_{t-1}+\lambda_2z_{t-2}+u_t$$

λ_1 and λ_2 being obtained from the equations

$$r_1=\lambda_1+\lambda_2r_1$$
$$r_2=\lambda_1r_1+\lambda_2$$

Returning to the first order autoregression model, often referred to as a Markov process, the r_1 values are obtained from the formula

$$r_1 = \left(\sum_{t=2}^{N} z_t z_{t-1} \right) \Big/ \left(\sum_{t=1}^{N} z_{t-1}^2 \right)$$

The variance of the z_t sequence is computed from

$$s_{ra}^2 = (1/N) \sum_{t=1}^{N} z_t^2$$

which is not unbiased. Actually, 12 degrees of freedom are lost when subtracting the seasonal variation and the mean. The unbiased variance should therefore be

$$\hat{s}_{ra}^2 = s_{ra}^2 (60/48)$$

A residual variance can now be defined by

$$\hat{s}_r^2 = \hat{s}_{ra}^2 (1 - r_1^2)$$

being the variance of the uncorrelated sequence u_t.

Table 6.3 summarizes the data on variances for the variables at Norrsjön and Kroksillret. It is seen that the autocorrelation is somewhat higher at Kroksillret except for sulphate, chloride and sodium. The higher values of r are due to the lake reservoir upstream which damps the fluctuations, removing more of the 'noise' (the $\hat{s}_{ra}^2$ part) than at Norrsjön.

The difference in variance $\hat{s}^2$ and $\hat{s}_{ra}^2$ reflects the significance of the seasonal variation. When the difference is small the seasonal pattern is part of the 'noise'. This part will be elaborated further in the next subsection where an analysis of variance will be performed.

As to the autocorrelation, this indicates that some of the information in the data is redundant. Part of the information in the data has already been used. This will have an effect on the significance tests of yearly means which are made up of twelve consecutive sets of data but not on the monthly means where data are spaced one year apart. Consider the yearly means. The variance of these will be the expected value

$$\mathrm{Var}(\bar{x}_j) = E\left\{ \left((1/12) \sum_{i=1}^{12} (x_{ij})^2 \right) \right\}$$

Table 6.3 Data from variances (total $\hat{s}^2$, random s^2_{ra} and residual $\hat{s}^2_r$) and on autocorrelation r_1

Property	Mean	$\hat{s}^2$	$\hat{s}^2_{ra}$	r_1	$\hat{s}^2_r$
I Kroksillret					
pH	6.71	0.0720	0.0842	0.69	0.0441
Alkalinity (meq l^{-1})	0.219	0.0013	0.0012	0.64	0.0007
Electrical conductivity (μS cm^{-1})	38.4	46.58	45.21	0.63	27.27
Sulphate (mg l^{-1})	4.23	0.2456	0.1777	0.13	0.1747
Chloride (mg l^{-1})	0.81	0.0438	0.0442	0.10	0.0438
Sodium (mg l^{-1})	1.27	0.0192	0.0141	0.13	0.0139
Potassium (mg l^{-1})	0.50	0.0116	0.0110	0.06	0.0110
Magnesium (mg l^{-1})	1.17	0.0326	0.0323	0.70	0.0165
Calcium (mg l^{-1})	4.41	0.4607	0.5048	0.52	0.3683
II Norrsjön					
pH	6.53	0.1097	0.1145	0.44	0.0923
Alkalinity (meq l^{-1})	0.163	0.0018	0.0013	0.28	0.0012
Electrical conductivity (μS cm^{-1})	32.6	42.42	22.76	0.53	16.37
Sulphate (mg l^{-1})	3.51	0.3405	0.3078	0.54	0.2180
Chloride (mg l^{-1})	0.63	0.0315	0.0209	0.49	0.0159
Sodium (mg l^{-1})	1.08	0.0198	0.0069	0.28	0.0064
Potassium (mg l^{-1})	0.30	0.0305	0.0304	0.13	0.0299
Magnesium (mg l^{-1})	0.89	0.0383	0.0313	0.47	0.0244
Calcium (mg l^{-1})	3.89	0.6166	0.4451	0.66	0.2512

which will be the sum of all variance and co-variance elements in a twelve by twelve matrix divided by 144. When the autocorrelation is zero only the diagonal elements contribute and the result is $\hat{s}^2_{ra}/12$ as expected for the mean error. For the simple autoregression model the sum of the matrix elements can be approximated by $(\hat{s}^2_{ra}/12)(1+r_1)/(1-r_1)$. One can consequently define an effective number of months in a year as $12(1-r)/(1+r)$. For Mg at Kroksillret the table shows an r_1 value of 0.7 which means an effective number of months in a year of 2.1. Since the variance is 0.0326, the squared mean error of yearly means becomes 0.0326/2.1 and the mean error 0.125 instead of 0.052 if r_1 had been zero.

Analysis of variance
This is a procedure very much in use since it forms an easy basis for testing significance of various groupings within data sets. In the present case data can be arranged to show yearly means and monthly means. The idea behind the analysis of variance is best perceived by carrying out the actual analysis on the data which have been used.

Suppose one wants to assess the significance of differences between years and between months. The data are arranged in a matrix of five rows and twelve columns. The square of deviations from the overall mean is then worked out for every element and summed to obtain the sum of squares of the total, labelled *SST*. The total variance is then obtained by dividing the *SST* by 59, the number of degrees of freedom. This would be the same as the variance $\hat{s}^2$ listed in Table 6.3. However, the variance as such is not of much interest in the analysis of variance procedure, only the sum of squares of 'total', *SST*.

Summing the rows, one obtains quantities which are twelve times the yearly means of each row. Consider a procedure by which all the single data in a row is made up of the mean of the row. Applying the same procedure as for the *SST* on the matrix, under these conditions, will give a sum of squares which reflects the differences between yearly means on a scale comparable to the *SST*. This new sum of squares 'between years' can be labelled *SSR*. It will certainly be smaller than the *SST*. Actually, dividing *SSR* by the number of years minus one – the proper degrees of freedom – gives the variance 'between years', i.e. how large the year to year variance is.

A similar procedure is now applied to the columns, i.e. the columns are all made up of monthly means. The sum of squares worked out with this arrangement can be labelled *SSC*, being the sum of squares 'between months'. Dividing this by 11, the proper number of degrees of freedom, gives the variance of the monthly mean values.

Returning now to the sum of squares, the *SST* represents the total variation within the data set. Part of this variation is accounted for by *SSR*, the sum of squares 'between years' and another part by *SSC* the sum of squares 'between months'. There is something left over which can be called *SSI* defined by

$$SSI = SST - SSR - SSC$$

being the part of sum squares which cannot be explained. One can refer to this part as the random part of sum of squares, or the sum of squares of 'random'. The degrees of freedom corresponding to this are then the 'total', 59, minus the 'between years', 4, minus 'between months', 11. This becomes 44. The variance of the random part is then $SSI/44$ and would correspond to $\hat{s}_{ra}^2$ of Table 6.3 save for the fact that the 'between years' variance is included in $\hat{s}_{ra}^2$. Adding *SSI* and *SSR* using this as the 'random' part with 48 degrees of freedom should give exactly the variances $\hat{s}_{ra}^2$.

The results are now entered into a table constructed as in Table 6.4. The

Table 6.4 Analysis of variance of pH and sulphate concentrations at Norrsjön and Kroksillret.

Source of variation	Symbol	Sum of squares	DOF	Variance	F-value
Norrsjön, pH					
Total	SST	6.4733	59		
Between years	SSR	2.0917	4	0.5229	6.66*
Between months	SSC	0.9293	11	0.0845	1.08
Random	SSI	3.4523	44	0.0785	
Norrsjön, sulphate					
Total	SST	20.0891	59		
Between years	SSR	2.6797	4	0.6698	2.44
Between months	SSC	5.3135	11	0.4830	1.76
Random	SSI	12.0959	44	0.2749	
Kroksillret, pH					
Total	SST	4.2540	59		
Between years	SSR	2.0607	4	0.5152	11.45*
Between months	SSC	0.2140	11	0.0195	2.30
Random	SSI	1.9793	44	0.0450	
Kroksillret, sulphate					
Total	SST	14.4805	59		
Between years	SSR	0.7925	4	0.1981	1.127
Between months	SSC	5.9528	11	0.5412	3.078
Random	SSI	7.7352	44	0.1758	

first column of the table lists the sources of variation corresponding in this case to 'total', 'between years', 'between months' and 'random'. In the next column the corresponding sum of squares are entered and in the next the number of degrees of freedom.

Dividing the sum of squares by the degrees of freedom gives the variances in the next column. The last column contains ratios of the variances 'between years' and 'between months' to 'random'. In the null hypothesis it is assumed at the outset that all variances come from the same statistical population and that there are no significant differences between them. The distribution of variance ratios, called F-values, for different sample sizes have been worked out and are found in many handbooks in tables as critical F-values at selected probability levels. When a computed F-value exceeds the critical F-value in the tables then the null hypothesis must be rejected.

The actual computational work in the analysis of variance is simplified if the general average is computed and subtracted from every data in the table under study. The SST is then obtained by squaring and summing all the squares. The SSR is obtained by summing the rows, squaring the sums, summing them and dividing by 12, the number of data in each sum. For the columns the same procedure is applied; summing the columns, squaring the sums and summing them and dividing the result by 5. This gives the SSC. Any computer program doing this work will be simple in any computer language. There are now all the items needed to complete the table on the results.

Some examples on the analysis of variance are shown in Table 6.4. The F-values marked with an asterisk are larger than the 5% critical F-value for corresponding degrees of freedom. It looks from the table as if there are significant differences in yearly means of pH at both Norrsjön and Kroksillret. However, the analysis of variance presupposes that the 'random' part is uncorrelated. This is not the case and the significance of differences between years is therefore doubtful. The application of analysis of variance to time sequences must be made with care.

6.2 Chemical budgets of basins

Budgets of chemical components in a basin are of great interest ecologically, geochemically and hydrologically. Because of varying climatic conditions it is necessary to continue such investigations for several years to give the results a kind of generality and also to be able to elucidate the influence of climatic elements on the budget, particularly on input from the atmosphere and discharge into the streams.

Part of budget investigations concerns the uptake by plant and release by decomposing litter of those elements which are intimately involved in what has been called mineral cycling through vegetation. This part is usually the concern of biologists. The input from the atmosphere is, however, a hydrochemical problem as well as discharge into the streams draining the basins.

There are the usual technical problems of sampling and budget calculations which need considerable attention before the results are reliable. Such problems and other associated problems will be discussed in the following subsections.

6.2.1 *Sampling problems*

Sampling of rainfall for chemical analysis is nowadays a well standardized procedure. The representativeness of one station, carefully selected, is in general sufficient for small basins unless there are large differences in altitude within the basin in which case a couple of sampling stations are needed. The sampling sites should be shielded from strong winds as well as from obvious sources of atmospheric constituents. Collection can be carried out on a monthly basis by having a storage bottle connected to a plastic funnel which collects the precipitation. Sophisticated rain samplers for chemical analysis have lids which open automatically at the beginning of rain and close at the end. They require electric power. They effectively prevent dust and trash accumulating in the funnel during dry spells. In winter they can be made to function so that the snow sample collected is melted and kept in the storage bottle at a suitable temperature.

Standard rain gauges should always be set up as well in order to get the best possible estimate of rainfall or snowfall since they usually are constructed in such a way as to minimize the errors caused by strong winds. When using these, the amounts of chemical constituents deposited during a month are worked out from the precipitation amount read from the standard gauge and the concentrations obtained by analysing the sample. During winter, in temperate climates snow surveys should be made to obtain the best possible figure of snowfall. Snowfall measured by standard raingauges can be beset with considerable errors.

The deposition obtained from the measurements is generally referred to as wet deposition. This implies that dry deposition of chemical constituents from the atmosphere also takes place. It certainly does but there is at present no reliable way to measure dry deposition in the same way as wet deposition. Dry deposition depends much on the nature of the surface cover – the type of vegetation, for instance – and also on the nature of the chemical constituents, be they gases or particles. There are some rough ways to guess the dry deposition using the concept of a vertical transport velocity in the air layer near the ground.

If this is known then the dry deposition will be the product of this velocity and the concentration of the chemical constituents in air (at the level for which the vertical transport velocity is valid) provided that the concentration of the same constituent is zero at the surface. The last condition may be true for gases like ammonia and sulphur dioxide but may fail for particulate matter. Vertical transport velocities can be estimated from humidity gradients and evapotranspiration rates. There are thus possibilities but they require additional measurements. The dry deposition can, however, also be estimated from budget computations using chloride as a tracer. This will be discussed later. Dry deposition is at most of the same order of magnitude as wet deposition.

Sampling of discharging water is done in the stream draining the basin at the point of exit. At this point, water discharge is also measured. Water discharge is usually obtained by recording the water level in a section using a calibration curve for the relation discharge-water level. The discharge rate can thus be computed for any interval of time and at any time. Time resolution of discharge is therefore usually adequate as long as water levels are recorded graphically or otherwise.

6.2.2 Budget calculations

Chemical assessment has to be made on samples collected at intervals. Too small intervals will make the cost prohibitively large. It is therefore necessary to plan the sampling intervals so that acceptable results are obtained. The flux of a substance, $F(t)$, is defined by

$$F(t) = Q(t)C(t)$$

where $Q(t)$ is the water discharge and $C(t)$ the concentration of the substance at time t. If $Q(t)$ and $C(t)$ are recorded continuously then one would obtain the monthly flux of the substance by integrating the product $Q(t)C(t)$ over a month. Using indexed variables the daily flux can be written

$$F_j = Q_j C_j$$

where Q_j is the discharge during day j, and C_j is the average concentration that same day. A monthly mean flux is then obtained as

$$\bar{F} = (1/M) \sum_{j=1}^{M} Q_j C_j$$

where M is the number of days of the month. Writing now

$$(1/M) \sum_{j=1}^{M} Q_j C_j = \overline{QC} = \bar{Q}\bar{C} + \overline{Q'C'} = \bar{F}$$

where the overbars indicate monthly means of the variables. The mean flux, $\bar{F}$ is equal to the mean discharge rate times the mean concentration, plus the mean of the products of deviations of Q and C from the means. The $\overline{Q'C'}$ can also be written $r s_Q s_C$ where r is the correlation coefficient between Q and C while s_Q and s_C are standard deviations of Q and C respectively. Thus as long as there is a correlation between discharge and concentration the product of the means will always be biased. Often the correlation between discharge and concentration is negative and in that case the product $\bar{Q}\bar{C}$ will give too high a value for $\bar{F}$. It seems therefore necessary under any conditions to get an estimate of $\overline{Q'C'}$, particularly of r the correlation coefficient.

In practice, samples for chemical analysis are taken at comparatively long intervals while the discharge is recorded continuously. Suppose one sample a month is taken (on the 15th) for analysis and the result is denoted C_m. Then an estimate of F is given by

$$\bar{F} = \bar{Q} C_m$$

$\bar{F}$ will be biased because the $\overline{Q'C'}$ term is ignored. If $\overline{Q'C'}$ is added to $\bar{F}$ the sum will be unbiased. $\bar{F}$ is the type of estimate which is generally used in budget computations for the practical reason that the costs of chemical analyses are very high compared to the cost of recording water levels for discharge computations. Thus, estimates based on $\bar{F}$ alone are likely to be too high. There is an unbiased estimate $F_m = Q_m C_m$ where m indicates a sample of discharge and concentration in the middle of the month, but this type of estimate does not seem to be used at all.

The effect of neglecting correlation was demonstrated on data collected on sediment transport in Swedish rivers by Nilsson (1971). Because suspended load in rivers are sensitive to discharge, Nilsson used fairly short sampling intervals from one or two days during spring floods to a week or more during winter. He determined also electrical conductivity on all samples and this can be used to illuminate the sampling problem. His data permit calculation of correlation coefficients between discharge and electrical conductivity in months when the sampling frequency is large. Four rivers were selected and for each month the biased flux $\bar{F}$ was calculated from the mean discharge and the conductivity of a sample collected on a day in the middle of the month. Adding the product $r s_Q s_C$ to this gave a new estimate, $\hat{F}$ which was unbiased. The results are shown in Table 6.5 with correlation coefficients and standard deviations. In the last

147

Table 6.5 Estimate of salt transports during 1969 in four Swedish rivers using different approaches*

River	Month	n	r	$\bar{F}$	$\hat{F}$	$\check{F}$
Viskan	1	7	0.36	2562	2608	4141
	2	3	–	3305	–	4445
	3	2	–	–		2854
	4	13	−0.79	4019	3932	5029
	5	15	−0.56	5033	4735	5641
	6	12	−0.75	3105	2836	3805
	7	9	−0.42	1023	964	884
	8	8	–	703	–	848
	9	9	–	1308	–	1076
	10	9	–	1300	–	1543
	11	8	–	7427	–	6959
	12	5	–	2652	–	3256
	Mean			3149	3019	3900
Klarälven	1	8	0.39	1382	1385	1379
	2	8	0.81	1682	1693	1619
	3	9	0.48	1788	1791	1657
	4	6	−0.81	2806	2709	2722
	5	9	−0.64	4423	4357	4547
	6	7	−0.93	2563	2457	2583
	7	5	0.33	1359	1367	1405
	8	8	0.56	1383	1401	1383
	9	9	0.27	1180	1184	1174
	10	9	0.08	1791	1792	1785
	11	8	−0.61	1552	1537	1541
	12	8	−0.41	1191	1187	1260
	Mean			1925	1905	1921
Ljusnan	1	2	–	62	–	63
	2	2	–	57	–	57
	3	2	–	52	–	54
	4	2	–	69	–	60
	5	20	−0.91	743	648	825
	6	27	−0.82	360	316	479
	7	8	−0.93	286	268	299
	8	9	−0.65	218	216	211
	9	9	−0.67	270	241	333
	10	9	−0.88	316	297	293
	11	8	−0.95	179	177	174
	12	6	−0.98	108	107	114
	Mean			311	284	341

Table 6.5 (continued)

River	Month	n	r	$\bar{F}$	$\hat{F}$	$\tilde{F}$
Ammerån	1	3	–	635	–	617
	2	3	–	604	–	607
	3	3	–	469	–	469
	4	4	–	712	–	869
	5	24	−0.84	7269	6795	7324
	6	16	−0.82	3274	3220	3390
	7	8	−0.81	1168	1143	1083
	8	8	−0.31	779	778	784
	9	9	−0.77	800	798	774
	10	9	−0.01	1578	1578	1410
	11	8	0.02	1279	1279	1290
	12	9	0.67	884	887	876
	Mean			2129	2096	2116

*n = number of samples during the month; r = correlation coefficient between electrical conductivity and discharge; $\bar{F}$ = monthly mean salt flow worked out from the monthly mean discharge and the electrical conductivity of one sample taken in the middle of the month; $\hat{F}$ = the same as $\bar{F}$, with addition of $rs_Q s_c$ (unbiased estimate); $\tilde{F}$ = mean of the products QC of the samples taken during the month.

column the flux $\tilde{F}$ obtained when using all the conductivity samples is given for comparison.

The results show nicely the bias in $\bar{F}$ but the error is sometimes fairly small. In river Klarälven the yearly mean fluxes of $\bar{F}$ and $\hat{F}$ are about the same, but in River Viskan the bias is about 5%. In general a bias is present when chemical sampling is made only once a month although the effect depends on upstream reservoirs. If large lakes are present the bias will be small.

Summing up, at present most estimates of the flux of chemical substances out of a basin are biased since they ignore the correlation between discharge and concentration. The bias term can be estimated at least roughly by recording electrical conductivity, together with streamflow which would enable the correlation coefficient between discharge and concentration to be established. Looking at the cluster description in Fig. 6.1 it is feasible to apply such a procedure to the single components of at least the geochemical group. The variances of the chemical components are obtained from the chemical data and the variance of the discharge from the records.

There is another point which needs attention when studying budgets in small basins. If the basin is situated close to a water divide of a much larger basin it is possible that part of the basin is the recharge area for a part of the regional or intermediate flow system. This means that this recharge never appears in the discharge. The stream discharge may be too small as compared with the groundwater recharge. The basin is thus not representative of the larger basin. Budget calculations will most likely be too low. It is difficult to give any figure for the possible error but the fact that streams can run dry during dry spells in temperate climates indicates at least intermediate groundwater flow which bypasses the discharge point. The smaller the basin, the larger the possible error.

If a small basin is selected at low altitude in a large basin the reverse may happen. The stream discharge is proportionately too high because of intermediate groundwater flow discharge. In addition the water may be too much mineralized in comparison to the small basin topography. The best place for a small basin ('small', meaning much less than 1 km^2) would be in mid-range of a larger basin. But the best way is, of course, to choose a larger basin, say a few km^2 in area.

Budget calculations also require estimates of inputs. Wet deposition can be measured but dry deposition is, as discussed, difficult to estimate. One way hinted at earlier is to use the chloride as a tracer. If there is no chloride source in the soils and rocks of the basin (as is often the case and can always be assessed) then the mean outflow of chloride has to be balanced by total deposition. Consequently, a factor can be worked out which, when multiplied by wet deposition is equal to the outflow of chloride. Assuming that the same factor can be applied to other constituents the best possible estimate of deposition rates is obtained. Some of them may still be biased because the mechanisms of dry deposition can differ for different elements. Sulphur, for example, can be absorbed by vegetation as sulphur dioxide and impinged on vegetation as particles whereas only the particulate sulphur will show up in the wet deposition. Consequently, the sulphur deposition may still be underestimated by the procedure and the same holds true for nitrogen compounds where ammonia as a gas can be absorbed by vegetation. The chloride ratio method as the suggested procedure can be called, is well founded for constituents in particulate form. A considerable part of sulphur in the atmosphere is present as sulphate particles and the method would thus work well. Near large sources of sulphur dioxide, deposition rates worked out by this method are likely to be underestimates.

An example of budget calculations is given in Table 6.6 where wet deposition, outflows and balances are listed. The area is one of the so-called representative basins studied in Sweden as part of the programme for the International Hydrological Decade. The basin is Kassjöån, situated in a forested area in central Sweden on granite–gneiss rocks. Wet deposition

Table 6.6 Depositions, discharges and balances of chemical substances in three sub-basins of the representative basin, Kassjöån, data given in mg m^{-2} year^{-1} (from Andersson-Calles and Eriksson (1979))

Source	S	Cl	NO_3-N	NH_3-N	Na	K	Mg	Ca
Wet deposition	902	267	229	210	221	124	67	435
Balances								
I Lilla Tivsjön								
Deposition	723	214	184	168	177	99	54	349
Discharge	402	214	22	9	364	113	340	1307
Balance	−321	0	−162	−159	187	14	286	958
II Öraåtjärn								
Deposition	787	233	200	183	193	108	58	380
Discharge	440	233	16	11	385	148	364	1344
Balance	−347	0	−184	−172	192	40	306	964
III Styggbergsbäcken								
Deposition	750	222	190	175	184	103	56	362
Discharge	353	222	13	10	414	119	285	2078
Balance	−397	0	−177	−165	230	16	229	1716

was measured in monthly collection of rainfall and snow and the deposition figures averaged over the years 1969–74 are shown in the table. The total deposition is calculated using the chloride ratio method just described. The flow of various components is calculated from the monthly water discharge and the concentrations in samples taken once a month. The bias in the outflow figures is judged as being small because the measuring points were mostly downstream of fair-sized lakes thus damping fluctuations in the concentrations.

Looking at the data for the three sub-basins listed in the table it is noted that dry deposition in this area is probably only a small fraction of the total deposition. Actually it looks as if chloride was retained in the basins but this discrepancy is not very serious. The most remarkable is the negative balance of sulphur, apparently being retained in the basins perhaps as organic sulphur storage or being re-emitted to the atmosphere as volatile compounds. The same experience is reported by Rosén (1982) in a study of forested basins not far from the Kassjöån area. Investigations in other representative basins in Sweden also show the negative balance of sulphur.

Also nitrogen compounds seem to disappear on the way to the outlet of the basins. Nitrogen compounds are incorporated into organic matter and

Principles and applications of hydrochemistry

parts of them may be lost in denitrification as nitrogen gas or as nitrous oxide. Parts of them may increase the storage in response to the increase in airborne nitrogen compounds noted during the last 30 years.

Budget calculations for basins on igneous rocks can give interesting information on mineral transformations. Table 6.7 shows calculated balances for all the sub-basins of the representative basins in Sweden of sodium, potassium, magnesium and calcium. The loss of potassium from the basins is comparatively small indicating that it is active in the formation of secondary minerals, presumably illites. There are some regional differences. The Velen basin is situated in the southwestern part of central Sweden, while Lappträsket is far to the north. But the rocks are all igneous, strongly fractured in some cases and the weathering going on seems fairly normal

Table 6.7 Balances of cations in the sub-basins of three representative basins in Sweden, data given in $mg\ m^{-2}\ year^{-1}$ (from Andersson-Calles and Eriksson (1979))

Basin and sub-basin	Na	K	Mg	Ca	Total
I The Velen basin					
1 Nolsjön	137	19	103	419	678
2 Velen	140	61	136	466	803
3 Sänningen	87	14	69	276	446
II The Kassjöån basin					
1 Lilla Tivsjön	187	14	286	958	1445
2 Eltnäs	199	7	181	687	1014
3 Öraåtjärn	192	40	306	964	1502
4 Kroksillret	153	34	254	840	1281
5 Styggbergsbäcken	230	16	229	1716	2191
6 Kassjöån	206	36	195	915	1352
7 Storsillret	192	36	236	940	1404
III The Lappträsket basin					
1 Tjärdalsselet	80	71	135	385	671
2 Norriåselet	108	86	144	433	771
3 Livastorpet	132	100	153	502	887
4 Spikforsen	128	91	170	494	883
5 Blåkölsbäcken	209	96	154	393	852
6 Lombergsfallet	142	116	137	414	809
7 Solmyren	116	120	211	720	1167
8 Vuoddasbäcken	166	62	151	459	838
9 Ytterholmen	122	96	151	427	796

when compared to results of similar studies in New Hampshire by Johnson *et al.* (1968). It is, of course, possible to calculate the amounts of various minerals needed to account for the balances.

6.2.3 *Information from chloride budget studies*

It seems of some interest to explore in detail the usefulness of budget studies for hydrological purposes. The fundamental element to consider is chloride for at least two reasons. Chloride behaves very much like water as studies using tritiated water have shown. Chloride is, from this point of view, a nearly perfect tracer for water. Secondly, igneous rocks and old sedimentary rocks are very low in chloride. The value obtained by Behne (1953) is 0.02% in all kinds of rocks: this value should be compared to the sodium concentration in igneous rocks which is close to 3%. Most natural waters have a chloride to sodium ratio of the order 1. The deposition rate of chloride is uncertain because the dry deposition cannot be measured directly. For calculation of balances this is partly overcome by the use of chloride, which is assumed to balance. The dry deposition of chloride is probably entirely due to impingement of sea salt particles on leaves and twigs. The impingement process is well studied and physically well understood. Coniferous trees like spruce and pine should be most effective in catching sea salt particles as well as any other particles from moving air. The impingement efficiency is thus related to the type of vegetation growing and to the wind condition and finally to the content of sea salt particles in the air. Studies of budgets of chloride will give good indirect information on dry deposition and the role of vegetation in this respect. However, more systematic studies are needed, particularly in warmer countries where data on deposition are so far few and scattered. As will be seen later, total deposition rates of chloride are needed if chloride is to be used as a tracer to calculate groundwater recharge. Thus, chloride budgets will aid such studies.

Before leaving this short section the relation between mineralization and hydrology should be mentioned. The present admittedly tentative indications from hard rock areas are that mineralization of groundwater proceeds at a slow but practically invariable rate. This rate is most likely diffusion-limited and so it has the character of zero order reactions. It is understandable in so far as primary minerals are rapidly hydrolysed but form a protective shell which then sets the speed. Thus, a high mineralization of a ground water indicates a long time of contact between water and minerals; this indicates a considerable storage or perhaps more precisely, a long turnover time. It is interesting to note that the transit time distribution of water is immaterial in this case, only the turnover time and

the contact surface is important. For very old groundwater, the mineralization may only result in more or less complete transformation of primary
minerals, the released cations being used again to build up clay minerals.
However, all primary minerals seem to have liquid inclusions of brine,
though in small concentrations. However, if they are released they may
accumulate in the water and reach considerable concentrations at least for
water of an age of several ten thousands of years. This tentative idea was put
forward by Nordstrom (1983).

6.3   Use of chloride and environmental isotopes in groundwater investigations

The properties of chloride as a tracer for water were discussed in the
previous section. Stable isotopes of water are also excellent because they
behave like water. Their use depends on fractionation of the isotopes. Once
a ratio of the isotopes is established it is a conservative property, as long as
the water is confined below ground level. To use the degree of fractionation
for assessing evaporation from surface bodies is so far not possible since the
precision of such estimates is far below that of conventional methods.
Chloride, however, is perfect in this respect. The fractionation on
evaporation from open water bodies is 100%. Also in evapotranspiration
processes chloride fractionation is complete, whereas for stable isotopes
there is no fractionation effect of evapotranspiration from soils and
vegetation.

6.3.1   *Calculation of groundwater recharge rates from chloride concentrations in groundwater*

Considering the mass flow of water in an aquifer in terms of, say,
$m^3 m^{-1} s^{-1}$ this refers to the flow rate of water integrated vertically from
the surface of the aquifer to the bottom. This is the usual approach used in
two-dimensional modelling of groundwater flow with either the finite
difference method (using grid systems of points) or the finite element
method (using vertical projections of surface elements). The mass balance
condition for these methods can be written

$$\partial F_x/\partial x + \partial F_y/\partial y = R$$

where F_x and F_y are the two flow components and R is the recharge rate
(when R is positive) or discharge rate (when R is negative). The equation
assumes a steady state of flow.

The flow of chloride can now be related to that of water by writing it as
$F_x C$ and $F_y C$ where C is the concentration of chloride averaged over depth.

154

Then the mass balance of chloride is

$$\partial(F_xC)/\partial x + \partial(F_yC)/\partial y = D$$

where D is the deposition rate (or discharge rate) of chloride. Considering the previous equation a combination of the two gives

$$F_x\partial C/\partial x + F_y\partial C/\partial y + CR = D$$

When salinity gradients are weak the approximation

$$R = D/C$$

can be used as the equation for groundwater recharge, in its dependence on the rate of deposition of chloride and the concentration of chloride in groundwater. In this form it will be explored further. However, the complete expression would read

$$R = (D - F_x\partial C/\partial x - F_y\partial C/\partial y)/C$$

In an aquifer, F_x and F_y certainly increase in the direction of flow but there is no particular reason for a gradient in concentrations of chloride unless there are systematic variations of the recharge rate and the deposition rate over the area. There may be random fluctuation in these over the area, in which case the gradients will average zero and the equation becomes

$$\bar{R} = \bar{D}\overline{(1/C)}$$

the overbars meaning averages over the area. The equation above implies that the mean deposition rate multiplied by the mean of inverted chloride concentrations gives the mean groundwater recharge. It is important to note that it is the inverted concentrations which have to be averaged not the concentrations. This means that high concentrations of chloride indicate very low recharge rates and they contribute very little when averaging $1/C$.

In an aquifer the sampling for calculating the average of $1/C$ should be restricted to the recharge areas. In a groundwater discharge area R is actually negative and the application of the equation fails.

The distribution of chloride concentrations appears to be lognormal. This is of considerable interest and justifies some comment. Consider a volume of groundwater, V, with a chloride concentration, C. Then the mass of chloride in this volume is $M = VC$. Now if the chloride should originate from the aquifer, distributed randomly, the volume V will increase its concentration of chloride by picking up chloride in the aquifer. This is the classical concept of the origin of chloride in groundwater. Considering the relation $M = VC$ the volume is then regarded as constant in time so the change in chloride concentration must be balanced by a change in mass. In other words

$$V \, \delta C = \delta M$$

where δ is an operator meaning a very small change. If now the additions occur randomly, i.e. δM is random, then according to the Central Limit Theorem, M should be normally distributed and, consequently so should C. One would under these circumstances expect a normal distribution of concentrations. This applies also to other constituents like bicarbonate when calcite is randomly distributed in an aquifer. An example of such a distribution is shown in Fig. 6.14. It is somewhat skew but this can be

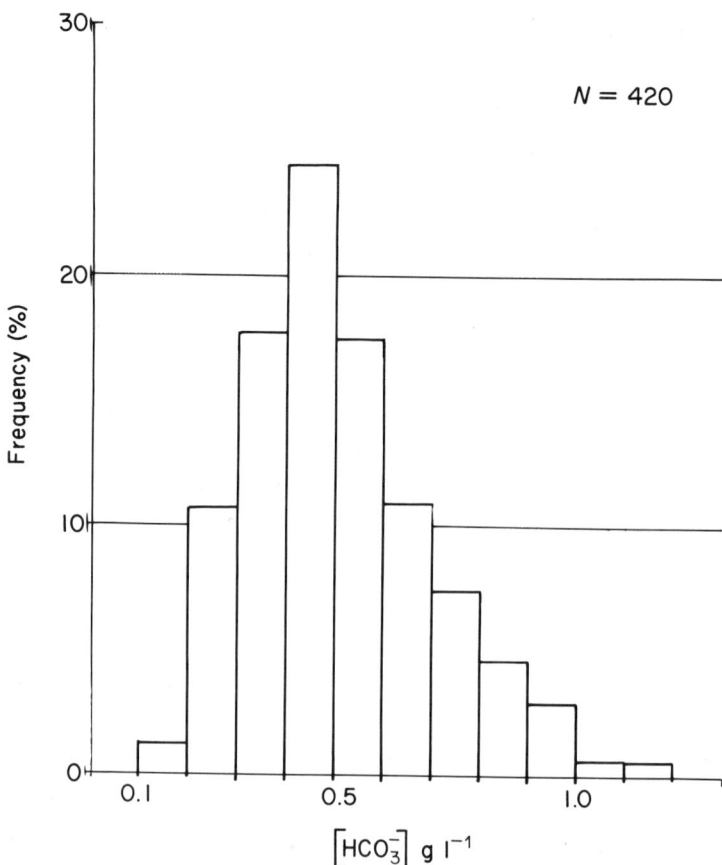

Fig. 6.14 The frequency distribution of bicarbonate in groundwater in the Delhi region in India, (from Eriksson (1976), reproduced by permission from the IAEA, Vienna).

explained as being due to evaporation effects. But it is certainly not log-normal. For chloride the distribution from the same area, the Delhi region, is shown in Fig. 6.15 using logarithmically spaced class intervals and it appears that this distribution is log-normal. It does not conform to the theoretical arguments presented and the origin of chloride is most unlikely to be the aquifer itself. If chloride is added by deposition only, then the mass M should be regarded as constant while the volume V will vary due to water extraction by evapotranspiration. Thus the balance for small changes must be written

$$C\ \delta V + V\ \delta C = 0$$

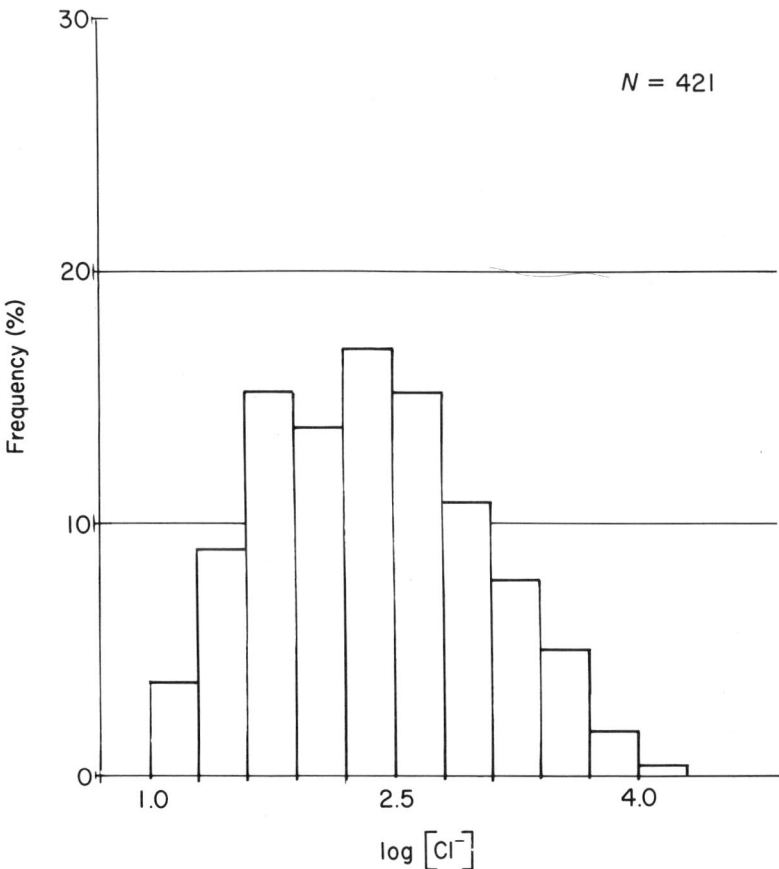

Fig. 6.15 The frequency distribution of chloride in groundwater in the Delhi region, (from Eriksson (1976), reproduced by permission from the IAEA, Vienna).

and

$$\delta C/C = -\delta V/V$$

Addition of all the small changes will then give

$$\ln (C/C_0) = -\ln (V/V_0)$$

where index 0 refers to the initial state. Thus if $\ln (V/V_0)$ is normally distributed then also $\ln (C/C_0)$ will be normally distributed. The criteria for the distribution of V are not obvious. One can give $\ln (V/V_0)$ a thermodynamic interpretation; that is, the energy spent on reducing the volume of water or one can argue that all changes in V are always proportional to V, in which case V will become log-normally distributed. There seems to be no need at present to dwell on this problem further. The fact is, that chloride concentrations in groundwater seem to be log-normally distributed. As another example, Fig. 6.16 again displays the lognormal features of groundwaters in Botswana convincingly. When the concentration C of chloride is log-normally distributed then also $1/C$ is log-normally distributed. This is understandable since $\ln (C) = -\ln (1/C)$.

Some examples on the use of the chloride balance for calculating groundwater recharges will now be given. In the Delhi region in India, a study of the hydrochemistry of groundwaters was made by Sett (1965) of the Geological Survey of India. More than 400 samples of groundwater were collected and analysed for, among other things, chloride. The frequency distributions in Figs 6.14 and 6.15 are constructed from these data. Eriksson (1976) used the chloride data together with some studies on chloride deposition from the atmosphere to work out the average groundwater recharge. The deposition of airborne chloride was estimated to be $3000 \text{ mg m}^{-2} \text{ year}^{-1}$ which includes also chloride in human consumption which will mainly be deposited on the soils. Since the harmonic mean of chloride is about 90 mg l^{-1} the mean annual recharge of groundwater in this region should be 3000/90, i.e. 33 mm. Considering the average yearly rainfall which is 500 mm, the figure 33 mm year^{-1} is about 7% of the yearly rainfall. Other measurements in the area carried out using tritium tracing of water in the intermediate zone, (Datta *et al.*, 1973) gave appreciably higher recharge values, about 200 mm year^{-1}. However, this is only one year whereas the chloride balance method give a long term average. The figure 33 mm year^{-1} may, however be low because some samples in recharge areas may have been included. On the other hand data on the peaks of environmental tritium in the intermediate zone investigated at several places in Northern Gujarat, thus fairly close to the Delhi region, gave recharge values between 15 and 63 mm year^{-1}, thus more in agreement with the chloride balance data.

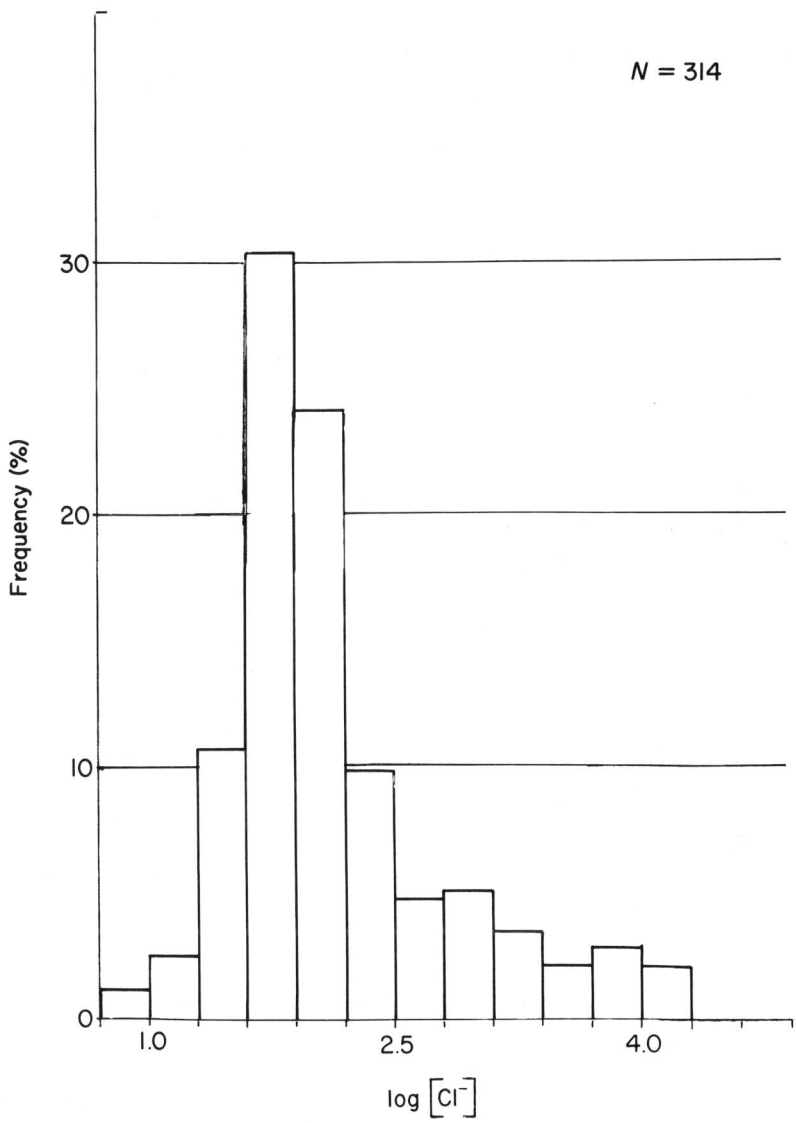

Fig. 6.16 The frequency distribution of chloride concentrations in groundwaters in the Kalahari part of Botswana.

In a study made by the Bureau de Recherches Geologiques et Minieres in Senegal, samples of rainfall and groundwater were collected in two areas near the coast, Tiaroye and Malika. The data are reported as a work of the Comite Inter-Africain d'Etudes Hydrauliques but the report carries neither the name of the author nor any date of printing.* In Tiaroye the chloride concentration in rainfall was found to be 11 mg l^{-1} while the harmonic mean of fourteen groundwater analyses gives 59 mg l^{-1}. The rainfall is 513 mm year^{-1} so the yearly groundwater recharge should be 96 mm year^{-1}. There may be some dry deposition of sea salt so near to the coast and if it is taken to be the same as that in rainfall, the yearly recharge figure doubles, i.e. becomes 192 mm year^{-1}. This should be compared to estimates based on the fluctuations of the groundwater level in response to rainfall which in 1963 was calculated to be 143 mm. Other estimates based on the evapotranspiration from soil as compared to rainfall gave a recharge figure of 159 mm for 1966.

Malika (which is close to the coastline) showed 20 mg l^{-1} as chloride in rain. Five wells gave a harmonic mean of 131 mg l^{-1} and hence, the recharge should be 78 mm year^{-1} if dry deposition is ignored; if dry deposition is considered, the recharge is probably about double this figure.

Data on chloride in groundwater near Fort Lamy in the Republic of Chad have a harmonic mean of 6.5 mg l^{-1}. There are no measurements on deposition rates in this area but measurements on wet deposition in the interior of Central Africa are consistent and show deposition rates from 250–450 mg m^{-2} year^{-1} (Eriksson, 1960). The total deposition rates may be higher but they are likely to decrease towards the north. A reasonable guess for Fort Lamy is 300 mg m^{-2} year^{-1} – perhaps it is even lower – which means that the groundwater recharge in this area would be $300/65 = 46$ mm year^{-1}.

In Tanzania some data were obtained from the local offices of the Water Department. One set is from a fairly arid part of Tanzania south of Arusha in the Kiteta district. The rocks are all igneous with sometimes a fairly thin soil cover. The chemistry of samples from 42 wells had been investigated with a harmonic mean for chloride of 77 mg l^{-1}. The deposition in this area was taken to be 300 mg m^{-2} year^{-1} so that the mean groundwater recharge becomes close to 4 mm year^{-1}. This is indeed a small recharge rate. In the Dodoma area, rather abundant groundwater is obtained at a system of wells at Makutupora, north of the city. The harmonic mean for chloride in the wells (ten analyses were made) had a value of 93 mg l^{-1} which is rather high. The deposition rate of chloride in this place is

*The report is available from: Comité Inter-African d'Etudes Hydrauliques, Bureau Téchnique, 25 Square Max Hymans, Paris XVe.

considered to be higher than in the Kiteta district and is set at 400 mg m^{-2} year^{-1}. This means a yearly groundwater recharge of about 4 mm year^{-1}.

In Botswana, the major aquifers are found in the Karoo formation of continental origin, now being covered by varying thicknesses of the Kalahari Sands. The area is arid with seasonal rains. Data have been collected by the Geological Survey Department in Lobatse, Republic of Botswana for a number of years and were made available in connection with a government supported pre-study of water resources in Botswana. The data were arranged according to localities and geological formation. Using a value of 500 mg m^{-2} year^{-1} for total deposition (in nearby Pretoria, 330 mg m^{-2} year^{-1} was obtained as wet deposition, Eriksson, 1960) the following ranges of local mean values of groundwater recharge were calculated.

Formation	Recharge (mm year^{-1})	Remarks
Ecca, Karoo	9–12	Mean values
Cave, Karoo	3–10	Mean values
Basalts, Karoo	9–10	Only a few values
Unidentified, Karoo	3–11	Mean values
Waterberg, Karoo	3–72	Single values
Granite, Karoo	7–72	Single values

There are a few others of interest where calculations are based on single values

Formation	Recharge (mm year^{-1})
Kata River	45
Dubwe	2
Francistown	32
Serowe, nine wells	9–70
Ghanzi, four wells	1–10
Lobatse, nine wells	3–130

There is a considerable variation in the results. The principal factor which determines the groundwater recharge rates in this region is the depth of the Kalahari sands. Whenever this blanket of sands is more than about 4 m deep, the groundwater recharge is considered to be practically nil (Foster *et al.* (1982)). The high recharge values are actually found in rock outcrops which are highly fissured such as the dolomite near Lobatse where recharge is calculated to be more than 100 mm year^{-1}.

The examples given are all from areas where information on deposition of chloride is either missing or incomplete. A considerable uncertainty, perhaps up to 100% can therefore not be avoided. Still the data calculated are certainly of interest and of guidance for water resources assessment in an area.

6.3.2 *Computation of groundwater recharge from hydrochemical data and groundwater flow patterns*

When the information on deposition rates and chloride concentrations in an aquifer is extensive, another procedure can be followed which gives a fairly detailed information on the distribution of groundwater recharge. One such case was described by Eriksson and Khunakasem (1969) and will be described in some detail. The area of interest is the Coastal Plain Aquifer in Israel. The groundwater conditions in this area were studied by the British in the period 1933–35 and the results are summarized in Figs 6.17 and 6.18. Deposition of chloride in rainfall was measured later by Yaalon and Katz (1962) and the result is shown in Fig. 6.19, again as a contour map with lines of equal deposition rates. Dry deposition was assumed to add another 30%, so the total deposition was obtained by multiplying the wet depositions by 1.3.

From the maps shown it is obvious that a more detailed procedure for calculation of groundwater recharge rates is likely to give information on the spatial variation of recharge. It is seen that the groundwater flow will be practically parallel towards the coast. Hence, considering the salt balance equation this can be written

$$d(F(s)C(s))/ds = D(s)$$

where the distance s is taken along streamlines which are supposed to be parallel. The deposition rate $D(s)$ is taken to be a function of s the distance from the boundary. Integrating the equation gives

$$F(s)C(s) = \int_0^s D(s')\,ds'$$

The integration was done graphically for a number of distances along each streamline. From the integral, the flow rate of water $F(s)$, is then obtained by dividing the integral by the concentration, C. The recharge rates are then

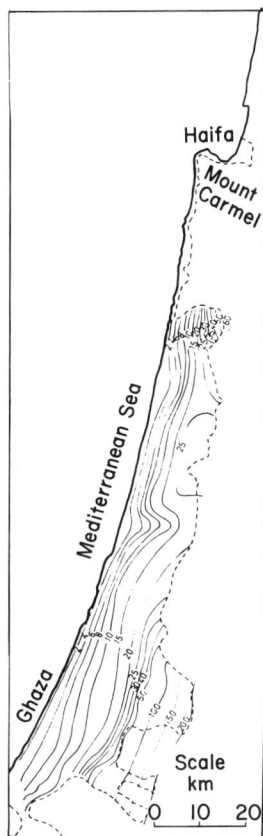

Fig. 6.17 Contour map of the water table in the Coastal Plain Aquifer in Israel for the years 1933–35, (from Eriksson and Khunakasem (1969), reproduced by permission from Elsevier Scientific Publishing Company).

obtained by differentiating $F(s)$ along s. In this way, a network of points with R values is obtained from which the distribution map in Fig. 6.20 is constructed. It is interesting to note the fairly large variation in recharge rates obtained. The largest ones are found near the coastline and are situated in sand dune areas which apparently are excellent for groundwater recharge.

The method described can also be used when streamlines are not parallel. The difference in treatment is that in this case the integration of deposition is made along bands bounded by streamlines and the groundwater flow as the total between the streamlines. In this way convergence and divergence of streamlines can be accounted for.

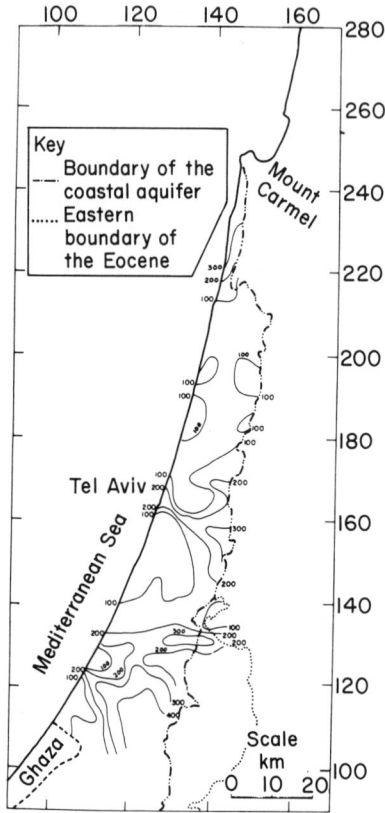

Fig. 6.18 Contour map of chloride concentrations (salinity) in mg l^{-1}, in the Coastal Plain Aquifer in Israel in the years 1933–35, (from Eriksson and Khunakasem (1969), reproduced by permission from Elsevier Scientific Publishing Company).

The method just described requires a considerable amount of information but it also makes good use of this information.

6.3.3 Interpretation of hydrochemistry patterns in groundwater in semi-arid climates

The mass balance of water and chloride discussed in the previous subsection can very well be used to explain or even predict salinity patterns of groundwater in semi-arid climates. Old continental rocks are strongly denuded and often have the typical flat expanses of alluvial – colluvial material of sometimes considerable thickness through which so called

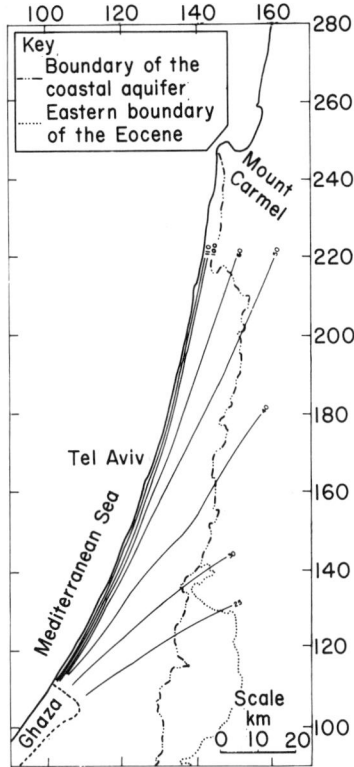

Fig. 6.19 Contour map of deposition of chloride in rainfall in kg ha^{-1} over the Coastal Plain Aquifer in Israel, (from Eriksson and Khunakasem (1969), reproduced by permission from Elsevier Scientific Publishing Company).

inselbergs protrude, looking like emerging mountain peaks. The flat expanses do undulate and may even show slight depressions. Because of very low sloping angles the groundwater levels are relatively close to the soil. A considerable part of the groundwater is recharged at the foot of inselbergs and in highly fractured rocks this groundwater finds speedy passages through the fractures to the loose deposits. A great deal of the recharged water will be lost by evapotranspiration in the depressions where salt also accumulates. Recharge of groundwater through loose deposits is nil wherever these are deep. Only in shallow parts can infiltration into fracture systems take place with any efficiency.

There will often be very large differences in salinity between recharge areas and discharge areas. The high salinity of the latter should not be taken as an indication that high salinity groundwater emerges. It is more likely

165

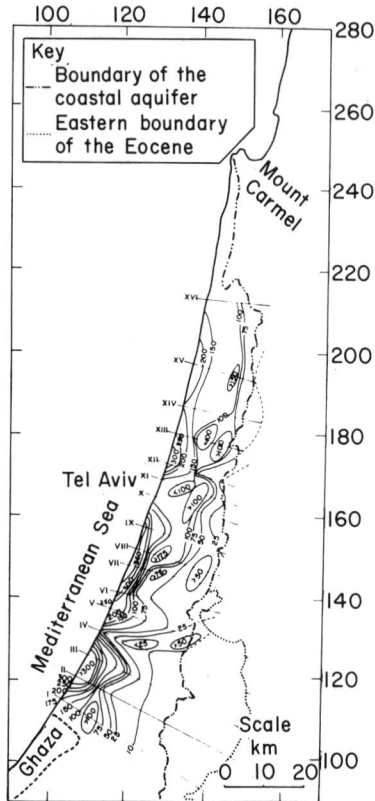

Fig. 6.20 Contour map of groundwater recharge rates in mm year^{-1} in the Coastal Plain Aquifer in Israel, (from Eriksson and Khunakasem (1969), reproduced by permission from Elsevier Scientific Publishing Company).

that the pressure head is strong enough to force groundwater through loose deposits towards the surface where it evaporates, leaving salts as residues. In fact, the emerging water may be of low salinity. High salinity of very shallow groundwater is just a sign of groundwater discharge. When deeper groundwater is saline this should be taken as emanating from a discharge area, i.e. downstream of a groundwater 'mound' of discharging water.

In areas like those described the least saline water is often found at some depth with saline water on the top. Again this is due to enrichment through evaporation in discharge areas and outflow from there as a layer that will float on top of the fresher water. Vertical mixing by convection in porous media does not seem to take place.

166

6.3.4 ^{14}C-dating of groundwater

The fundamental physics and chemistry of ^{14}C was treated in a previous section. Its application to groundwater dating has its limitations as was pointed out. With good sedimentary rocks of marine origin the problems of interpretation should be small.

The sampling technique is somewhat laborious. In order to obtain enough inorganic carbon for ^{14}C counting, 100–200 litres of water are needed. Since this is expensive to transport to the laboratory, inorganic carbon is extracted in the field by acidifying the water and washing out the carbon dioxide with a stream of pure nitrogen gas which passes through a carbon dioxide absorbant. The absorbant is then sent to the laboratory for further processing. A sample of water is, however also collected for chemical and ^{13}C analyses.

Whenever there is an uncertainty about the correction needed for age calculations one should collect samples in the flow direction, since the difference in ^{14}C ages between such samples is not affected by an unknown fraction of 'dead' carbon, as long as this part is constant. The age differences are interpreted as travelling times between the points of sampling. From this the turnover time of the groundwater and the recharge can be assessed provided the geometry of the aquifer is known.

^{14}C analyses should be consistent with the chemistry of the water and with the stable isotope concentrations. With this information at hand the best possible interpretation is achieved. In hard rock areas with calcite-coated fractures one should be careful when interpreting ^{14}C data; not even horizontal differences in age can be relied upon.

^{14}C dating of groundwater can give information on the transit time distribution since this is simply related to the age distribution within the aquifer (Eriksson, 1961). The prerequisite for this is, however, that no large scale mixing has taken place in the aquifer. This requires great care in sampling and use of packers is inevitable in order to get as unmixed samples as possible. The sampling should be representative of the entire water mass in the aquifer. Investigations aiming at assessments of transit time distributions will be costly, but from many points of view they will be extremely valuable for hydrochemical modelling of aquifers. The age distribution is constructed from the age determinations, weighting the samples according to the volumes they represent. If this distribution is denoted $M(\tau)$ being defined as the fraction of water with an age equal to or less than τ then the transit time distribution $F(\tau)$ is given by

$$F(\tau) = 1 - dM(\tau)/d\tau$$

where $F(\tau)$ is the fraction of transit times equal to or less than τ. The success of such a procedure depends entirely on the quality of determined ages.

6.3.5   *The origins of groundwater as derived from hydrochemistry and stable isotopes*

In any formation groundwater may have different origins. In alluvial formations there are two possible origins: rainfall and infiltration from rivers, provided that there is a sink somewhere in the aquifer, such as evaporation from depressions. It is likely that rainfall and river water have different ^{18}O content: in a long river, enough evaporation may take place to increase the ^{18}O; on the other hand, rainfall which has an intensity large enough to cause groundwater recharge, is often depleted in ^{18}O. With this knowledge it should be possible to work out the fractions of the two sources at any point in the aquifer, sometimes an important piece of information from a planning point of view. Chloride concentration is another possible tracer if rainfall and river water differ. However, one should be aware of the fact that groundwater discharge areas may have higher concentrations of chloride due to evapotranspiration. This does not affect ^{18}O because it requires a surface reservoir to be enriched.

Considering three possible sources of groundwater, two tracers are needed to obtain an unambiguous estimate of the fractions of each origin. For four sources three tracers are needed. As a general rule if there are n different sources of water one needs $n-1$ independent tracers to calculate the fraction of water from each source. ^{18}O and deuterium are not independent so one of them is of no use. But chloride and ^{18}O *are* independent and can be used to calculate the three fractions of different water present in any sample. Thus when rainwater, river water and sea water are the sources, then ^{18}O and chloride are suitable, provided of course, that there is a difference in ^{18}O between the rainfall and river water. Nitrate can be used as a third tracer permitting determination of the fractions of water from four different sources. However, all this requires that concentration levels are not increased by evapotranspiration. As long as samples are drawn from recharge areas there should be no objections to the use of the three tracers.

In some cases the ^{18}O deuterium relation can be used as another tracer, i.e. both ^{18}O and deuterium are useful. In general this requires processes which would change the slope of the $\delta D - \delta^{18}O$ line from the standard value of 8 (as shown for one of the sources of groundwater).

The way to work out the fractions is simple. Consider three different sources of groundwater. The concentrations of tracer number one are denoted by α_1, α_2 and α_3 and of tracer number two by β_1, β_2 and β_3. In a sample, the fractions of each source are x, y and $1-x-y$ and the tracer concentrations found by analysis are α and β. Then the mass balance of tracers is set by

$$x\alpha_1 + y\alpha_2 + (1-x-y)\alpha_3 = \alpha$$
$$x\beta_1 + y\beta_2 + (1-x-y)\beta_3 = \beta$$

from which x and y are obtained. The equations can also be written

$$x(\alpha_1 - \alpha_3) + y(\alpha_2 - \alpha_3) = \alpha - \alpha_3$$
$$x(\beta_1 - \beta_3) + y(\beta_2 - \beta_3) = \beta - \beta_3$$

It can be seen from this that it is still possible to use the relations if one of the tracers has the same concentration in two of the three sources, provided that the other tracer has different concentrations in the two sources. It can also be seen that if two tracers co-vary exactly then the two equations become identical; i.e. there is in effect only one equation and the problem cannot be solved.

6.3.6 Study of the run-off process using environmental isotopes

There are quite a number of studies where environmental isotopes are used to distinguish between two sources of water in a stream, usually between rainfall or snow-melt and local groundwater. The problem is thus a special case of those discussed in the previous subsection. It is a problem of two sources and one tracer. The outcome of all these analyses has changed the traditional picture of storm run-off as consisting of a major fraction of rainwater and a small fraction of groundwater, so-called base flow. The isotope studies show convincingly that storm run-off consists to a considerable degree of groundwater that is being pushed out due to groundwater recharge. The results are, of course, relevant for hydrochemical modelling. One could most likely extend these kind of investigations to several sources of water in a stream if they could be distinguished in the basin.

6.4 The effects of acid deposition

During the last few decades it has become evident that atmospheric depositions in Europe and in parts of North America are very acid, this acidity being generally associated with sulphate and nitrate. The source of this acidity is the combustion of sulphur-containing fuels like coal and oil and the fixation of atmospheric nitrogen during combustion; the latter is caused by the strong heating and rapid cooling of stack gases and exhaust gases from internal combustion engines.

The effects of deposition of acid substances on soil and water were first noticed in South Norway, due to the gradual disappearance of salmon species from rivers. Later, they were noted in Sweden by the appearance of clear water lakes which had become devoid of life except for Sphagnum mosses which grew on the bottom of the lakes. The pH of the water was around 4 or less.

The acid rainfall in Scandinavia initiated considerable research in Norway and Sweden, a research which also aimed at the more fundamental problem of the hydrogen ion budget in soils, i.e. the production of hydrogen ions, the annihilation of hydrogen ions, the inflow and outflow of hydrogen ions. This is an interesting approach; deposition of acid from the atmosphere is just one factor in the hydrogen budget, a factor which may be important in some cases and unimportant in others. However, there may also be other ways of describing the influence of acid rainfall.

6.4.1 *The hydrogen ion budget of the vegetation–root zone system in groundwater recharge areas*

Common cations such as K^+, Na^+, Mg^{2+} and Ca^{2+} are taken up by plant roots and transported through the plant in company with organic acids. In plant material there is generally an excess of cations over anions of mineral acid. Hence, on root uptake the metal cations in the ion exchange complex in the soil are replaced by hydrogen ions from the roots in exchange for the metal ions. Thus, during plant growth, there is a production of hydrogen ions in the root zone due to this uptake of metal cations. However, there is some compensation for this: when nitrate is taken up by plant roots it is converted into amino groups and thus takes up hydrogen. And, similarly, when sulphate is converted into organic hydrogen sulphide groups.

When vegetation turns to 'litter' it decomposes and returns the metal cations to the soil. Initially, they are locked up in the organic matter but when it breaks down the cations are released, consuming the equivalent amount of hydrogen ions. Thus, the net amount of hydrogen ions produced in the past by root uptake of metal cations is, simply, equivalent to the storage of these cations in living plants. For grasslands, this amount is not particularly large when compared to the total storage in the root zone, as long as all vegetable matter is returned to the root zone as litter. In forests, the vegetative storage of metal cations is considerable. The amount of hydrogen ions produced in the past by root uptake may be of the same order of magnitude as the total storage of exchangeable ions in the root zone.

The leaching of metal cations from the root zone when accompanied by bicarbonate ions equals the hydrogen production in the root zone. This is a continuous process and in the long term, is more important than the effect of storage of metal cations in vegetable matter. Leaching of metal cations in the root zone is consequently very low when the pH of the water in the root zone is about 5.5, simply because the bicarbonate concentration is very low.

When plant material is removed from land to feed people and animals, for example, or trees chopped down for timber, metal cations are also removed. Hence, in most land-use practices, a net production of hydrogen ions will take place over a period of time because metal cations are removed, being frequently transported to urban sites and released into water courses and city dumps. Removal of cations with crops is similar to leaching.

Hydrogen ion concentrations in the root zone are regulated by the partial pressure of carbon dioxide and the bicarbonate concentration. Reaction with silicates produce bicarbonate and leaching removes it. There is an upper limit for the hydrogen ion concentration set by the partial pressure of carbon dioxide. Consequently, the influence of hydrogen ion production on the pH of soil water is limited.

6.4.2 *The metal cation balance in the vegetation–root zone system in groundwater recharge areas*

In terms of the metal cation balance of the root zone, there will be outflows through leaching and such land use practices as described which remove part of the vegetation from the land. There will be a production of metal cations through weathering of silicates in the root zone and there may be addition of metal cations by liming and fertilizer application. In the absence of such practices, the weathering will cover loss by leaching and removal of crops. The rate of leaching will depend not only on the rate of weathering, but also on the rate of removal of metal cations by crops. A stationary state (in the static sense) will always be reached.

The system described is interesting in that the loss of metal cations by leaching is likely to decrease when they are removed by cropping the land. But so also will the storage. On the other hand, the application of lime and fertilizers is expected to increase the loss by leaching.

6.4.3 *The influence of acid deposition on the metal cation balance of the root zone in groundwater recharge areas*

Distilled water in equilibrium with atmospheric carbon dioxide has a pH of about 5.6. In soil water, at a partial pressure of carbon dioxide about 100 times that in the atmosphere, the pH should be around 4.6, provided the bicarbonate concentration is equal to the hydrogen ion concentration. Normally, even under land use without liming, the pH of the water in the root zone is likely to be at least between 5 and 6. Acid deposition must then be great enough to produce a pH in the water of the root zone below 5. The acidity is due to nitric oxides, sulphur trioxide and hydrochloric acid. It is not immediately obvious from the chemical composition of rainfall which of these components are responsible for the acidity. The only thing apparent is that hydrogen ions are present in a concentration exceeding that expected from carbonic acid alone. The origin of the acidity is the concern of air chemists.

When acid rain reaches the root zone its acidity may have been partly neutralized by the leaching of litter on the ground, but this has no effect on the metal ion balance as such. In the root zone, the added hydrogen ions will reach equilibrium with the cation exchange complex, the metal cations being pushed into solution in exchange for hydrogen ions. This is, however, a transient state. A new equilibrium state is reached when the acid rain, after loss of water by evapotranspiration, is in complete equilibrium with the cation exchange complex. The time needed to reach this state depends on the turnover time of the metal cations stored in the root zone.

Thus in the transient state, the pH will be governed primarily by the original set-up of the cation exchange complex but in the stationary state, the pH will be determined by the deposition rate and the water balance (provided that nothing happens to the mineral acid anions, chloride, sulphate and nitrate). As to chloride, this is a conservative property and need not be considered further. Sulphate, however, can be converted into organic sulphide groups and stored, in which process an equivalent amount of hydrogen ions are consumed. Inorganic reactions resulting in hydrogen sulphide are unlikely in the root zone if it is well aerated. Sulphate ions can be sorbed by mineral surfaces or by amorphous hydroxides like aluminium hydroxide. Nordstrom, (1982) obtained good evidence that at least a couple of aluminium sulphate minerals may be present in soils below pH 5. These are jurbanite, $AlSO_4OH \cdot 5H_2O$ and alunite, $KAl_3(SO_4)_2(OH)_6$. The effect of the formation of these is two-fold: it reduces both sulphate concentrations and aluminium concentrations and in addition, consumes hydrogen ions. Nordstrom thinks that what has been described as adsorbtion of sulphate by soils is in fact the formation of basic aluminium sulphates.

The reaction involving jurbanite can be written

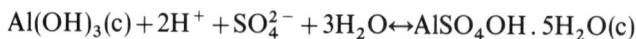

$$Al(OH)_3(c) + 2H^+ + SO_4^{2-} + 3H_2O \leftrightarrow AlSO_4OH \cdot 5H_2O(c)$$

from which it can be seen that the acidity is reduced in equivalent amounts to the reduction of sulphate ions. For alunite the reaction is

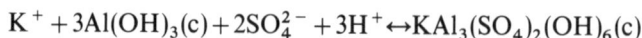

$$K^+ + 3Al(OH)_3(c) + 2SO_4^{2-} + 3H^+ \leftrightarrow KAl_3(SO_4)_2(OH)_6(c)$$

which shows that one mole of sulphate and 1.5 moles of hydrogen ions are removed simultaneously. Alunite is stable at a higher pH than jurbanite. It was pointed out earlier when discussing balances of chemical constituents, that a loss of sulphate was experienced when comparing input and outflow of sulphur. Since the present state (with respect to acid rainfall) is still transient one can hardly expect sulphate to balance, if it reacts with aluminium in the way described. Since there are considerable amounts of aluminium hydroxide in soils it will take some time before such a state is reached.

As to nitrate, there are a number of indications in Scandinavia at least, that it also is lost on the way from precipitation to outflow. However, this may only be typical for the temperate humid climate of coniferous forests.

Thus, when aluminium hydroxides are present there is a reduction of acidity in deposition by processes in the root zone and on the ground surface. The actual degree of reduction will depend on the initial concentrations of sulphate and hydrogen ions. Considering alunite, the pK of its formation is 85.4 according to Nordstrom. This gives the equation

$$p[K^+] + 3pAl + 2pSO_4 + 6pOH = 85.4$$

and for $p[K^+] = 5$ (0.4 mg l^{-1})

$$3pAl + 2pSO_4 - 6pH = -3.6$$

Combining this with the gibbsite equation

$$pAl + 3pOH = 33.9$$

one obtains

$$2pSO_4 + 3pH = 20.7$$

In the reaction three hydrogen ions are used up for every two sulphate ions. Considering this and the equilibrium equation above, the amount of sulphate 'fixed' can be calculated knowing the initial concentration and the initial pH. The following calculated values can be regarded as typical (SO_4 in mg l^{-1})

pH initially	3.7	4.0	3.7	4.0
pH at equilibrium	3.94	4.16	4.22	4.36
SO_4 initially	9.6	9.6	19.2	19.2
SO_4 at equilibrium	3.64	7.4	9.1	15.1

The reduction in sulphate is thus quite significant. Of particular interest, is the fact that a higher initial sulphate concentration brings about a larger reduction in hydrogen ion concentration than a lower one, even when the initial acidity is the same. Addition of lime to the soil would release the sulphate and the corresponding number of hydrogen ions.

6.4.4 *Acid precipitation and the hydrogen ion balance of basins*

The processes discussed so far have been limited to the root zone of groundwater recharge areas. However, sulphate sorption can certainly proceed also in the intermediate zone and in the saturated zone as long as there are clay minerals ready to release aluminium. The sulphate sorption would then proceed as a front in the direction of water movement.

Local groundwater flow systems will in general have a shallow water circulation. Equilibrium in the sulphate sorption would therefore be reached fairly soon with the full impact of deposited acidity. The intermediate and regional groundwater flow systems require much longer exposure to acid rain before a breakthrough of acidity occurs. At the same time, weathering of minerals supply hydroxyl ions which remove the hydrogen ions. At a given deposition rate, there is a certain fraction of the recharge area where weathering is sufficient to neutralize the acid and even to produce excess alkalinity. The problem can be considered in the following way: the production of alkalinity within a basin proceeds at a constant rate; when the acid deposition rate is greater than the alkalinity production rate, then in a stationary state, the outflow will be acidic, in other cases it will be alkaline. The effect of sulphate sorption is to delay the time when a completely stationary state is reached.

References

Behne, W. (1953) Untersuchungen zur Geochemie des Chlor und Brom, *Geochim. et Cosmochim. Acta* **3**, 186–214.

Datta, P. S., Goel, P. S., Rama, P. A. Sc. and Sengal, S. P. (1973) Groundwater recharge in western Uttar Pradesh, *Proc. Indian Acad. Sci.*, **78**, Ser A, No. 1, 1–12.

Davis, J. C. (1973) *Statistics and data analysis in geology*, John Wiley and Sons Inc., London.

Eriksson, E. (1960) The yearly circulation of chloride and sulfur in nature; meteorological, geochemical and pedological implications. Part II, *Tellus*, **12**, 63–109.

Eriksson, E. (1976) The distribution of salinity in groundwaters in the Delhi region and recharge rates of groundwater, in *Interpretation of environmental isotope and hydrochemistry data in groundwater hydrology*, IAEA, Vienna, pp. 171–7.

Eriksson, E. and Khunakasem, V. (1969) Chloride concentration in groundwater, recharge rates and rate of deposition of chloride in the Israel Coastal Plain, *Journal of Hydrology*, **7**, 178–97.

Falkenmark, M. (ed.) (1979) Hydrological Data – Norden, representative basins,

Kassjöån, Sweden, data 1966–74, Swedish National Committee for the IHP Stockholm.

Foster, S. S. D., Bath, A. H., Farr, J. L. and Lewis, W. J. (1982) The likelihood of active groundwater recharge in the Botswana Kalahari, *Journal of Hydrology*, **55**, 113–36.

Grip, H. (1982) Water chemistry and runoff in forest streams at Kloten, UNGI report No. 58, University of Uppsala

Johnson, N. M., Likens, G. E., Bormann, F. H. and Pierce, R. S. (1968) Rate of chemical weathering of silicate minerals in New Hampshire, *Geochim. et Cosmochim. Acta*, **32**, 531–45.

Nilsson, B. (1971) Sediment transports in Swedish rivers, UNGI Report No. 4, University of Uppsala.

Nordstrom, D. K. (1982) The effect of sulphate on aluminium concentrations in natural waters: some stability relations in the system $Al_2O_3-SO_3-H_2O$ at 298 K, *Geochim. et Cosmochim. Acta*, **46**, 681–92.

Nordstrom, D. K. (1983) Personal communication.

Rosén, K. (1982) Supply, loss and distribution of nutrients in three coniferous forest watersheds in central Sweden, Reports in Forest Ecology and Forest Soils, No. 41, Swedish University of Agricultural Sciences.

Sett, D. N. (1965) Groundwater geology of the Delhi region, *Bull. Geol. Survey of India*, Ser B, p. 123.

Yaalon, D. H. and Katz, A. (1962) The chemical composition of precipitation in Israel, *Proceedings, Fourth Congress, Israeli Association for the Advancement of Science*, p. 189–90.

Further reading

Andersson-Calles, U. M. and Eriksson, E. (1979) Mass balance of dissolved inorganic substances in three representative basins in Sweden, *Nordic Hydrology*, **10**, 99–114.

Eriksson, E. (1961) Natural reservoirs and their characteristics, *Geofisica International*, **1**, 27–43.

Yevjevich, V. (1972) *Stochastic processes in hydrology*, Water Resources Publications, Fort Collins, Colorado, USA.

Appendices

Appendix A Elements and compounds commonly occurring in natural waters, their chemical formula, atomic weight or molecular weight and the equivalent weights of ions

Substance	Formula	Atomic weight	Molecular weight	Equivalent weight
Aluminium	Al	26.98	–	–
Aluminium hydroxide	$Al(OH)_3$	–	78.00	26.00
Iron	Fe	55.85	–	–
Ferric ion	Fe^{3+}	–	55.85	18.62
Ferrous ion	Fe^{2+}	–	55.85	27.93
Fluor	F	19.00	–	–
Fluoride ion	F^-	–	19.00	19.00
Phosphorous	P	30.975	–	–
Phosphoric acid	H_3PO_4	–	97.999	–
Dihydrogen phosphate	$H_2PO_4^-$	–	96.991	96.991
Hydrogen phosphate	HPO_4^{2-}	–	95.983	47.992
Phosphate	PO_4^{3-}	–	94.975	31.658
Hydrogen	H	1.008	–	–
Hydrogen ion	H^+	–	1.008	1.008
Hydrogen gas	H_2	–	2.016	2.016
Calcium	Ca	40.08	–	–
Calcium ion	Ca^{2+}	–	40.08	20.04
Calcium carbonate	$CaCO_3$	–	100.09	50.05
Potassium	K	39.100	–	–
Potassium ion	K^+	–	39.100	39.100
Carbon	C	12.011	–	–
Carbon dioxide	CO_2	–	44.011	–
Carbonic acid	H_2CO_3	–	62.027	–
Bicarbonate ion	HCO_3	–	61.019	61.019
Carbonate	CO_3^{2-}	–	60.010	30.005

Appendix A (continued)

Substance	Formula	Atomic weight	Molecular weight	Equivalent weight
Silicium	Si	28.09	–	–
Silica	SiO_2	–	60.09	–
Silicic acid	H_4SiO_4	–	96.12	–
Chlorine	Cl	35.457	–	–
Chloride ion	Cl^-	–	35.457	35.457
Magnesium	Mg	24.32	–	–
Magnesium ion	Mg^{2+}	–	24.32	12.16
Magnesium carbonate	$MgCO_3$	–	84.33	42.17
Manganese	Mn	54.94	–	–
Manganese(II) ion	Mn^{2+}	–	54.94	27.47
Manganese dioxide	MnO_2	–	86.94	–
Permanganate ion	MnO_4^-	–	118.94	118.94
Sodium	Na	22.991	–	–
Sodium ion	Na^+	–	22.991	22.991
Nitrogen	N	14.008	–	–
Nitrogen gas	N_2	–	28.016	–
Ammonia	NH_3	–	17.032	–
Ammonium ion	NH_4^+	–	18.040	18.040
Nitrate ion	NO_3^-	–	62.008	62.008
Oxygen	O	16.000	–	–
Oxygen gas	O_2	–	32.000	–
Water	H_2O	–	18.016	–
Sulphur	S	32.066	–	–
Hydrogen sulphide	H_2S	–	34.082	–
Sulphur dioxide	SO_2	–	64.066	–
Sulphuric acid	H_2SO_4	–	98.082	–
Sulphate ion	SO_4^{2-}	–	96.066	48.033

Appendix B Gibb's standard free energies of formation, G, and standard enthalpies of formation, ΔH_f^0, for a selected number of constituents

Formula	Substance	ΔG_f^0 (kcal mol^{-1})	ΔH_f^0 (kcal mol^{-1})	Reference number*
Al(c)	Aluminium	0	0	1
Al^{3+}(aq)	Aluminium ion	-119.5	$-$	2
Al(OH)$_3$(c)	Gibbsite	-277.0	-306.8	2
AlO(OH)(c)	Diaspor	-221.2	-235.5	$-$
Al(OH)$_2^+$(aq)	Aluminium dihydroxy ion	$-$	$-$	$-$
C(c)	Carbon, graphite	0	0	1
CO$_2$(g)	Carbon dioxide, gas	-94.2598	-94.1	1
CO$_2$(aq)	Carbon dioxide, dissolved	-92.31	-98.7	1
H$_2$CO$_3$(aq)	Carbonic acid	-149.0	-167.0	1
HCO$_3^-$(aq)	Bicarbonate ion	-140.31	-165.2	1
CO$_3^{2-}$(aq)	Carbonate ion	-126.22	-161.6	1
C$_6$H$_{12}$O$_6$(aq)	Glucose	-217.02	$-$	1
Ca(c)	Calcium	0	0	1
Ca^{2+}(aq)	Calcium ion	-132.18	-129.8	1
CaCO$_3$(c)	Aragonite	-269.53	-288.5	1
CaCO$_3$(c)	Calcite	-269.89	-288.98	5
CaMg(CO$_3$)$_2$(c)	Dolomite	$-$	$-$	$-$
CaSO$_4$.2H$_2$O(c)	Gypsum	-429.19	-483.1	1
CaSO$_4$(c)	Anhydrite	-315.56	-342.4	1
Ca$_3$(PO$_4$)$_2$(c)	Calcium phosphate	-929.7	-986.2	1
CaHPO$_4$(c)	Calcium monohydrogen phosphate	-401.5	-435.2	1
Ca(H$_2$PO$_4$)$_2$(c)	Calcium dihydrogen phosphate	-672	-744.4	1
CaF$_2$(c)	Calcium fluoride	-280.08	-292.59	4
Cl$_2$(g)	Chlorine, gas	0	0	1
Cl$^-$(aq)	Chloride ion	-31.350	-40.0	1
HCl(g)	Hydrogen chloride, gas	-22.769	-22.1	1
HCl(aq)	Hydrogen chloride, dissolved	-31.350	-40.0	1
Fe(c)	Iron	0	0	1
Fe^{2+}(aq)	Ferrous ion	-20.30	-21.0	1
Fe^{3+}(aq)	Ferric ion	-2.53	-11.4	1
Fe(OH)$_2^+$(aq)	Ferridihydroxy ion	-106.2	$-$	1
Fe(OH)$^{2+}$(aq)	Ferrimonohydroxy ion	-55.91	-67.4	1
Fe(OH)$_2$(c)	Ferrous hydroxide	-115.57	-135.8	1
FeCO$_3$(c)	Siderite	-161.06	-178.0	1

Appendix B (continued)

Formula	Substance	ΔG_f^0 (kcal mol^{-1})	ΔH_f^0 (kcal mol^{-1})	Reference number*
FeS(c)	Troilite	−22.32	−22.72	–
FeS$_2$(c)	Pyrite	−39.84	−42.5	1
Fe(OH)$_3$(c)	Ferric hydroxide	−166.0	−197.0	1
FeO(OH)	Goethite	−117.0	–	2
Fe$_2$O$_3$(c)	Hematite	−177.7	−196.5	2
Fe$_3$O$_4$(c)	Magnetite	−242.4	−267.0	3
H$_2$(g)	Hydrogen gas	0	0	1
H$^+$(aq)	Hydrogen ion	0	0	1
Mg(c)	Magnesium	0	0	1
Mg^{2+}(aq)	Magnesium ion	−108.99	−110.3	2
Mg(OH)$_2$(c)	Brucite	−199.5	−221.0	2
MgCO$_3$(c)	Magnesite	−246.1	−266.1	3
Mn(c)	Manganese	0	0	1
Mn^{2+}(aq)	Manganous ion	−54.4	−53.3	1
Mn^{3+}(aq)	Manganic ion	−19.6	−27.0	1
Mn(OH)$_2$(c)	Pyrochroite	−146.9	−166.9	1
MnO(c)	Manganous oxide	−86.8	−92.0	1
MnS(c)	Alabandite	−53.3	-	1
Mn(OH)$_3$(c)	Manganic hydroxide	−181.0	−212.0	1
Mn$_2$O$_3$(c)	Manganic sesquioxide	−212.3	−232.1	1
Mn$_3$O$_4$(c)	Hausmannite	−306.0	–	1
MnO$_2$(c)	Pyrolusite	−111.1	−124.5	3
MnO$_4^-$(aq)	Permanganate ion	−107.4	−129.7	1
N$_2$(g)	Nitrogen gas	0	0	1
N$_2$O(g)	Nitrous oxide	24.76	19.5	1
NO(g)	Nitric oxide	20.719	21.6	1
NO$_2$(g)	Nitrogen dioxide	12.39	8.1	1
NH$_3$(g)	Ammonia, gas	−3.976	−11.0	1
NH$_3$(aq)	Ammonia, dissolved	−6.36	−19.3	1
NH$_4^+$(aq)	Ammonium ion	−19.0	−31.7	1
O$_2$(g)	Oxygen gas	0	0	1
H$_2$O(l)	Water, liquid	−56.690	−68.32	1
H$_2$O(g)	Water, vapour	−54.635	−57.798	1
OH$^-$(aq)	Hydroxyl ion	−37.595	−55.0	1
P(c)	Phosphorous	0	0	1
H$_3$PO$_4$(aq)	Phosphoric acid	−274.2	−308.2	1
H$_2$PO$_4^-$(aq)	Dihydrogen phosphate ion	−271.3	−311.3	1
HPO$_4^{2-}$(aq)	Monohydrogen phosphate ion	−261.5	−310.4	1

Appendix B (continued)

Formula	Substance	ΔG_f^0 (kcal mol^{-1})	ΔH_f^0 (kcal mol^{-1})	Reference number*
PO_4^{3-}(aq)	Phosphate ion	-245.1	-306.9	1
K(c)	Potassium	0	0	1
K^+(aq)	Potassium ion	-67.46	-60.2	2
Si(c)	Silicium	0	0	1
SiO_2(c)	Silica, quartz	-204.6	-217.6	2
SiO_2(am)	Silica, amorphous	-203.3	$-$	2
H_4SiO_4(c)	Silicic acid, orto	-312.5	-348.0	2
Na(c)	Sodium	0	0	1
Na^+(aq)	Sodium ion	-62.589	-57.2	1
NaCl(c)	Sodium chloride	-91.785	$-$	1
Na_2SO_4(c)	Sodium sulphate	-302.78	$-$	1
S(c)	Sulphur	0	0	1
SO_2(g)	Sulphur dioxide	-71.79	-70.8	1
SO_3(g)	Sulphur trioxide	-88.52	-94.5	1
SO_4^{2-}(aq)	Sulphate ion	-177.34	-216.9	1
H_2SO_4(aq)	Sulphuric acid, dissolved	-177.34	-216.9	1
H_2S(g)	Hydrogen sulphide, gaseous	-7.892	-4.8	1
H_2S(aq)	Hydrogen sulphide, dissolved	-6.54	-9.4	1
HS^-(aq)	Hydrogen sulphide ion	3.01	-4.2	1
S^{2-}(aq)	Sulphide ion	22.1	8.6	1

*References (see p. 31 for details)
1 Latimer (1956)
2 Tardy & Garrels (1974)
3 Handa (1975)
4 Nordstrom & Jenne (1977)
5 Jacobson & Langmuir (1974).

Appendix C   Chemical composition of river water at Norrsjön and Kroksillret

	Jan	Feb	Mar	Apr	May	June	July	Aug	Sept	Oct	Nov	Dec
Norrsjön												
pH												
1970	6.5	6.4	6.3	5.9	6.2	6.4	6.2	6.3	6.2	6.6	6.3	6.1
1971	6.2	6.2	6.3	6.3	5.9	6.9	6.3	6.4	6.5	6.7	6.5	6.6
1972	6.6	6.5	6.4	6.4	6.3	7.0	6.4	6.4	6.6	6.4	6.7	6.4
1973	6.3	6.4	6.6	6.7	6.5	6.6	7.1	7.0	7.0	7.1	7.1	7.3
1974	6.9	7.1	7.3	6.0	7.0	6.5	6.6	6.2	6.5	6.8	6.8	6.3
Alkalinity (μeq l^{-1})												
1970	181	189	222	227	79	100	114	129	143	162	140	111
1971	150	184	169	169	57	205	137	143	158	149	113	179
1972	234	210	192	166	153	121	137	120	142	162	171	183
1973	141	189	201	193	138	139	174	179	181	186	213	211
1974	221	296	240	109	141	191	195	100	127	142	142	128
Electrical conductivity (μS cm^{-1})												
1970	36	37	34	32	21	22	23	26	31	23	29	27
1971	30	39	34	36	21	35	28	28	29	29	31	36
1972	44	41	38	38	26	32	30	30	30	32	32	30
1973	34	38	37	50	27	31	34	32	28	30	32	33
1974	33	52	49	45	32	37	37	27	29	30	29	29
Sulphate												
1970	5.22	4.98	4.62	4.71	3.27	3.39	3.30	3.12	3.03	3.27	3.63	3.18
1971	3.42	3.24	3.18	3.93	2.64	3.12	3.03	3.03	3.06	3.06	3.48	3.39
1972	3.90	3.82	3.54	3.36	2.94	3.15	3.42	2.91	3.06	3.18	4.08	3.15
1973	3.66	3.87	3.54	3.09	3.18	3.30	3.60	4.08	4.35	4.53	3.48	3.42
1974	3.24	3.39	3.48	5.46	3.48	3.60	3.72	3.15	3.18	2.73	3.24	3.06
Chloride												
1970	0.96	0.89	1.10	0.88	0.41	0.46	0.58	0.53	0.47	0.89	0.90	0.61
1971	0.91	0.91	0.71	0.75	0.33	0.34	0.42	0.47	0.41	0.54	0.99	0.69
1972	0.87	0.80	0.74	0.64	0.51	0.41	0.60	0.46	0.46	0.54	0.51	0.51
1973	0.53	0.69	0.60	0.72	0.50	0.54	0.59	0.55	0.46	0.48	0.75	0.66
1974	0.56	0.41	0.70	0.75	0.43	0.56	0.72	0.71	0.75	0.66	0.66	0.71
Sodium												
1970	1.24	1.24	1.30	1.28	0.73	0.88	0.96	0.93	1.00	1.04	1.17	1.05
1971	1.22	1.29	1.15	1.23	0.74	0.92	0.94	1.00	1.02	1.04	1.06	1.16
1972	1.38	1.28	1.18	1.26	0.94	0.81	0.98	0.91	0.98	1.00	0.98	1.03
1973	1.04	1.22	1.25	1.10	0.90	1.04	1.02	1.01	1.08	1.09	1.21	1.22
1974	1.18	1.22	1.24	1.08	0.98	1.12	1.13	0.99	1.06	1.12	1.12	1.10

Appendix C (continued)

	Jan	Feb	Mar	Apr	May	June	July	Aug	Sept	Oct	Nov	Dec
Potassium												
1970	0.26	0.25	0.31	0.26	0.39	0.26	0.30	0.18	0.22	0.44	0.24	0.15
1971	0.23	0.58	0.25	0.49	0.33	0.29	0.29	0.29	0.29	0.30	1.32	0.28
1972	0.38	0.36	0.34	0.36	0.29	0.23	0.14	0.10	0.15	0.16	0.15	0.20
1973	0.23	0.22	0.23	0.28	0.23	0.23	0.28	0.26	0.25	0.27	0.30	0.26
1974	0.34	0.32	0.38	0.81	0.30	0.29	0.26	0.12	0.35	0.21	0.16	0.14
Magnesium												
1970	1.22	1.10	1.32	1.14	0.63	0.68	0.74	0.77	0.83	0.85	0.45	0.38
1971	0.43	0.90	0.92	1.26	0.70	0.75	0.73	0.78	0.80	0.85	0.89	0.92
1972	1.19	1.12	1.06	0.94	0.79	0.78	0.83	0.85	1.10	0.91	0.89	0.98
1973	0.83	1.00	1.02	0.90	0.77	0.79	0.94	0.96	0.95	0.96	0.98	0.98
1974	1.04	1.04	1.10	1.08	0.90	1.05	1.00	0.85	0.85	0.90	0.90	0.35
Calcium												
1970	4.85	5.00	5.27	5.10	2.55	3.10	3.36	3.84	3.88	3.84	3.70	3.36
1971	4.15	4.45	4.60	4.63	2.44	3.26	3.56	3.77	3.91	4.00	4.10	4.52
1972	5.55	5.31	5.07	4.77	3.92	3.55	3.83	3.75	3.90	3.99	4.32	4.45
1973	4.05	4.76	3.90	3.50	2.90	2.20	2.60	3.10	2.60	2.70	3.10	3.15
1974	3.00	3.00	4.68	4.16	2.86	4.54	4.31	3.69	4.15	4.28	4.22	4.22

Kroksillret

pH												
1970	6.7	6.5	6.6	6.4	6.5	6.5	6.6	6.5	6.4	6.5	6.5	6.6
1971	6.4	6.4	6.5	6.5	6.2	6.5	6.6	6.6	6.5	7.1	6.5	6.7
1972	6.7	6.4	6.4	6.6	6.4	6.9	6.6	6.8	6.8	6.6	6.7	6.6
1973	6.6	6.8	6.8	6.8	6.9	7.1	7.1	7.1	7.1	7.3	7.0	7.2
1974	6.8	7.0	7.4	7.2	7.2	6.7	6.6	6.6	6.8	6.8	6.9	6.5

Alkalinity (μeq l^{-1})												
1970	222	238	214	250	197	175	203	188	215	182	182	190
1971	186	203	184	207	158	180	193	189	226	197	196	199
1972	222	188	195	322	161	190	204	198	221	248	224	242
1973	199	228	230	207	192	201	260	296	331	271	252	245
1974	256	294	264	281	211	234	232	193	231	206	195	220

Electrical conductivity (μS cm^{-1})												
1970	38	39	35	30	31	30	31	31	25	26	26	37
1971	35	38	38	39	35	36	37	35	38	40	37	40
1972	44	40	38	48	34	36	38	36	34	42	38	37
1973	48	42	41	34	32	38	45	43	43	37	37	38
1974	39	66	55	56	41	43	44	38	39	35	35	41

Appendix C (continued)

	Jan	*Feb*	*Mar*	*Apr*	*May*	*June*	*July*	*Aug*	*Sept*	*Oct*	*Nov*	*Dec*
Sulphate												
1970	5.34	5.34	4.74	4.68	4.14	4.05	3.54	3.45	3.72	3.93	4.20	4.38
1971	4.68	4.38	4.32	4.47	4.02	3.78	3.78	3.66	3.69	3.96	4.05	4.38
1972	4.56	3.96	4.47	4.23	3.84	4.02	4.05	3.96	4.08	4.14	4.11	4.47
1973	4.89	4.53	4.59	4.14	4.23	3.90	4.14	4.29	4.77	5.02	4.17	4.35
1974	4.20	6.03	4.20	4.26	4.02	3.87	4.05	4.08	4.26	2.58	4.14	4.32
Chloride												
1970	0.96	0.85	0.88	0.57	0.69	0.74	0.92	0.72	0.61	0.66	0.82	0.80
1971	0.96	0.32	0.75	0.56	0.61	0.62	0.56	0.66	0.60	0.77	0.81	0.83
1972	0.92	0.69	0.60	0.64	0.83	0.55	1.29	0.69	0.83	0.86	0.79	0.64
1973	1.01	0.78	0.92	0.81	0.68	0.77	0.80	0.84	0.84	0.86	1.74	0.99
1974	0.80	1.24	1.01	0.92	0.72	0.89	0.81	0.89	0.85	0.89	0.71	1.22
Sodium												
1970	1.30	1.31	1.31	1.26	1.13	1.20	1.14	1.10	1.16	1.08	1.25	1.30
1971	1.36	1.36	1.30	1.29	1.16	1.08	1.16	1.15	1.23	1.24	1.24	1.28
1972	1.39	1.30	1.25	1.24	1.21	0.91	1.28	1.14	1.26	1.24	1.14	1.18
1973	1.50	1.37	1.47	1.28	1.17	1.23	1.24	1.34	1.54	1.30	1.33	1.36
1974	1.38	1.93	1.31	1.28	1.21	1.30	1.26	1.24	1.27	1.28	1.20	1.50
Potassium												
1970	0.53	0.47	0.49	0.45	0.45	0.54	0.46	0.43	0.44	0.44	0.65	0.46
1971	0.44	0.46	0.42	0.45	0.41	0.45	0.43	0.40	0.45	0.47	0.62	0.48
1972	0.56	0.47	0.47	0.47	0.62	0.40	0.59	0.35	0.48	0.55	0.41	0.42
1973	0.95	0.48	0.61	0.42	0.39	0.39	0.49	0.46	0.55	0.54	0.50	0.47
1974	0.68	0.91	0.59	0.59	0.49	0.55	0.51	0.46	0.48	0.48	0.47	0.66
Magnesium												
1970	1.29	1.45	1.32	1.29	1.10	1.00	1.08	1.00	1.08	1.06	0.60	0.59
1971	0.57	1.13	1.14	1.57	1.38	1.10	1.08	1.06	1.21	1.16	1.13	1.17
1972	1.32	1.23	1.21	1.16	1.02	1.02	1.09	1.11	1.11	1.16	1.09	1.23
1973	1.26	1.26	1.30	1.16	1.06	1.07	1.30	1.34	1.48	1.32	1.16	1.18
1974	1.26	1.44	1.28	1.26	1.25	1.25	1.20	1.15	1.15	1.10	1.20	1.25
Calcium												
1970	4.70	4.85	4.93	4.72	4.11	4.20	4.42	4.42	4.37	4.15	4.00	1.62
1971	4.74	4.70	4.92	5.16	4.20	4.17	4.55	4.40	4.98	4.70	4.65	4.73
1972	4.93	4.73	4.65	5.30	4.40	4.40	4.55	4.40	4.77	5.02	4.86	5.17
1973	4.90	4.97	4.10	3.70	3.60	2.70	3.40	4.10	3.70	3.20	3.80	3.25
1974	3.25	3.90	4.86	4.53	4.37	4.48	4.68	4.76	5.00	4.72	4.92	5.78

Index

Index